CARE FOR THE WORLD

Convening leading scholars to reflect on the practical and philosophical implications of religious values, this volume is an accessible introduction to Catholic social thought on contemporary affairs. Its gracefully written chapters cover three themes – direct environmental policy implications of *Laudato Si'*, philosophical alternatives to dominant policy discourse, and a renewed political economy based on robust conceptions of human flourishing. *Care for the World* offers learned reflections on what it would mean to express an ethic of compassion in an era of climate crisis.

Frank Pasquale is Professor of Law at the University of Maryland, and author of *The Black Box Society* (2015). He is an expert on the law of artificial intelligence and algorithms. He has been published in *Commonweal,* and his scholarly work has explored the implications of Catholic social thought for innovation policy.

LAW AND CHRISTIANITY

The Law and Christianity series publishes cutting-edge work on Catholic, Protestant, and Orthodox Christian contributions to public, private, penal, and procedural law and legal theory. The series aims to promote deep Christian reflection by leading scholars on the fundamentals of law and politics, to build further ecumenical legal understanding across Christian denominations, and to link and amplify the diverse and sometimes isolated Christian legal voices and visions at work in the academy. Works collected by the series include groundbreaking monographs, historical and thematic anthologies, and translations by leading scholars around the globe.

Books in the series

Under Caesar's Sword: How Christians Respond to Persecution
edited by Daniel Philpott and Timothy Samuel Shah

God and the Illegal Alien
Robert W. Heimburger

Christianity and Family Law
John Witte, Jr. and Gary S. Hauk

Christianity and Natural Law
Norman Doe

Agape, Justice, and Law
Robert F. Cochran, Jr and Zachary R. Calo

Great Christian Jurists in English History
Mark Hill, QC and R. H. Helmholz

Great Christian Jurists in Spanish History
Rafael Domingo and Javier Martinez-Torron

Calvin's Political Theology and the Public Engagement of the Church
Matthew J. Tuininga

God and the Secular Legal System
Rafael Domingo

How Marriage Became One of the Sacraments
Philip Reynolds

Christianity and Freedom (Volume I: Historical Perspectives, Volume II: Contemporary Perspectives)
edited by Timothy Samuel Shah and Allen D. Hertzke

The Distinctiveness of Religion in American Law
Kathleen A. Brady

Pope Benedict XVI's Legal Thought
Marta Cartabia and Andrea Simoncini

The Western Case for Monogamy Over Polygamy
John Witte, Jr.

Church Law in Modernity: Toward a Theory of Canon Law Between Nature and Culture
Judith Hahn

Care for the World: Laudato Si' and Catholic Social Thought in an Era of Climate Crisis
edited by Frank Pasquale

Care for the World

LAUDATO SI' AND CATHOLIC SOCIAL THOUGHT IN AN ERA OF CLIMATE CRISIS

Edited by

FRANK PASQUALE

University of Maryland

CAMBRIDGE
UNIVERSITY PRESS

CAMBRIDGE
UNIVERSITY PRESS

University Printing House, Cambridge CB2 8BS, United Kingdom

One Liberty Plaza, 20th Floor, New York, NY 10006, USA

477 Williamstown Road, Port Melbourne, VIC 3207, Australia

314–321, 3rd Floor, Plot 3, Splendor Forum, Jasola District Centre, New Delhi – 110025, India

79 Anson Road, #06–04/06, Singapore 079906

Cambridge University Press is part of the University of Cambridge.

It furthers the University's mission by disseminating knowledge in the pursuit of education, learning, and research at the highest international levels of excellence.

www.cambridge.org
Information on this title: www.cambridge.org/9781316510469
DOI: 10.1017/9781108227049

First published 2019

Printed and bound in Great Britain by Clays Ltd, Elcograf S.p.A.

A catalogue record for this publication is available from the British Library.

Library of Congress Cataloging-in-Publication Data
NAMES: Pasquale, Frank, editor.
TITLE: Care for the world : Laudato si' and Catholic social thought in an era of climate crisis / edited by Frank Pasquale.
DESCRIPTION: New York, NY : Cambridge University Press, 2018. | Includes bibliographical references.
IDENTIFIERS: LCCN 2018034109 | ISBN 9781316510469 (hardback)
SUBJECTS: LCSH: Catholic Church. Pope (2013– : Francis). Laudato si'. | Human ecology – Religious aspects – Catholic Church. | Ecotheology. | Climatic changes – Effect of human beings on. | Christian sociology – Catholic Church.
CLASSIFICATION: LCC BX1795.H82 C37 2018 | DDC 261.8/8–dc23
LC record available at https://lccn.loc.gov/2018034109

ISBN 978-1-316-51046-9 Hardback

Contents

Contributors

Anthony Annett is Climate and Sustainable Development Advisor at the Centre for Sustainable Development, Columbia University.

David L. Gregory is the Dorothy Day Professor of Law and Executive Director of the Center for Labor and Employment Law at St. John's University School of Law.

Vincent Ialenti is a Ph.D. Candidate in Anthropology at Cornell University and a MacArthur Nuclear Waste Solutions Fellow at The George Washington University's Elliott School of International Affairs.

Amanda Jaret is a staff attorney at the National Labor Relations Board.

Zachary Loeb is a Ph.D. Candidate in the History and Sociology of Science Department at the University of Pennsylvania.

Massimiliano Montini is Associate Professor of Law at the University of Siena.

Frank Pasquale is a Professor of Law at the University of Maryland.

Eduardo M. Peñalver is the Allan R. Tessler Dean at Cornell Law School.

Nathan Schneider is Assistant Professor of media studies at the University of Colorado Boulder.

Mark Shiffman is Chair of the Department of Humanities at Villanova University.

Alessandro Spina is a member of the Legal Service at the European Commission and Visiting Lecturer at the University of Fribourg.

Francesca Volpe is Research Fellow in Law at the University of Siena.

Acknowledgments

I first wish to thank Lawrence Joseph for suggesting the idea of this collection to me. I have benefitted from his stunningly eloquent poetry and his keen insights on politics, policy, and law. His intellect, and his capacity for hope (despite a remarkably clear-eyed grasp of the brokenness of the world), is rooted in a deep appreciation of the transcendent. That sensibility was an inspiration for me as I assembled and edited this collection.

As the collection took shape, Michael Perry offered sage advice and encouragement. John Witte has been a supportive series editor. The anonymous reviewers of the essays in this volume were both expert and encouraging, helping us to shape a more coherent set of statements on the enduring importance of *Laudato Si'*.

On a more personal note, I should acknowledge those who played a role in my own faith formation. The Catholic Church deeply shaped my parents' sense of life's ultimate meaning and purpose, and they in turn shaped mine. I acknowledge the great good fortune to have seen, in them, two of the many cultural styles of the faith: a demanding and metaphysical expression often found in Northern Europe, and the *joie de vivre* of Mediterranean Catholicism. In these two very different styles, they modeled for me the ways in which faith can both lend gravity to, and lighten the burden of, our journey through this world. They also did the hard work of getting me to First Communion and confirmation classes, and paying for my Catholic schooling when they could afford it. The Church, in turn, was a rare fixed point of stability in their lives, which were too often upended by what business gurus and neoliberal economists would approvingly deem "disruption" and "labor market mobility."

Catholic primary school in Kentucky and Oklahoma offered yet another experience of faith formation, and I thank all the teachers there (including especially Ruella Yates) for the discipline they imposed and the creativity they inspired.

At Harvard and Yale, I found several brilliant scholars who blended faith and reason. Father Bryan Hehir, despite so many academic and other obligations, was generous with his time with the Harvard Catholic Students Association, showing us the rigor of theology – as did Chaplain Jacqui Landry, who was an exceptionally warm and welcoming presence. Father Bob Beloin at Yale promoted "small church communities" that fostered genuine friendship and bible study among his congregants. Learned women and men at each institution helped me to "keep the faith" despite the many challenges to it raised by my reading in social theory.

While I taught at Seton Hall University (named after Saint Elizabeth Ann Seton), I was blessed to have many colleagues whose wisdom and learning reflected their faith. Father Nicholas Gengaro graced our community with homilies at weekday masses that reflected a deep appreciation of the relevance of scripture and Catholic social teaching to our daily lives, and our community. The faculty also exuded an inclusive ecumenism, with colleagues of all faiths (and none) comfortable discussing ultimate issues in wide-ranging and learned conversations. I wrote a tribute to one of them, Marc Poirier, when he passed away, and I still miss his work to bring opportunities for spiritual practice (in his case, grounded in Buddhist meditation) to a wider community.[1]

My friends Cat, Luanne, Mark, and Chris continue to exemplify both grace and theological learning for me, freely sharing their wide reading and experiences of worship and prayer. They are also inspirations.

For all that it has done for me and my community, any honest acknowledgment section for a book like this must also confront the fact that the Catholic Church is far from perfect, and itself has fallen short in many ways in my lifetime (and certainly before as well). For giving me the hope and faith that positive evolution in practice and doctrine is possible today, I thank public intellectuals like Father James Martin. And I thank them also for bringing the good news of the faith to political disputes over gun control and immigration policy. I also thank the Jesuits at "Pray as You Go" for helping to bring contact with Jesus into the lives of me and so many other Catholics. When this project or other research was trying, I found myself centered and at piece when listening to their digital breviary.

Finally, my thanks to the many dedicated people at Cambridge University Press who have helped to bring this book into being. It is an honor to be part of the Law and Christianity series.

1 Frank Pasquale, "The Scholarship of Resonance in an Era of Discord," *Seton Hall Law Review*, 47 (2017), 685.

Introduction

The Ethical, Legal, and Political Significance of *Laudato Si'*

Amanda Jaret and Frank Pasquale

Faith offers us many spiritual gifts. One of the greatest of these gifts is a sense of perspective – an appreciation of the smallness of our span of life on earth, and the greatness of eternity. A contemplative life enables us to perceive our surroundings *sub species aeternitatis* – a perspective transcending petty demands or ephemeral desires. Saints like Francis of Assisi and Teresa of Lisieux exemplify this gift of stepping outside of time, outside of human plans, to apprehend in radical simplicity the timeless virtues of self-giving, service, and contemplation of the divine.

Faith has also offered us another set of gifts: a sense of urgency in the face of injustice, and the bedrock principles to help us recognize such injustice. Catholic churches have fought segregation and have aided civil society campaigns against the structural sins of racism, exploitation, and authoritarianism. *The Catholic Worker* identified revolting violations of human rights and dignity, and fought to rectify them. For over a century, the Catholic Church's social encyclicals have applied biblical teaching and theological insight to critique dominant political and economic arrangements. Catholic social teaching has sparked moral concern and political action to make corporal acts of mercy at least in part a responsibility of the state.

One of the greatest tasks of spiritual discernment is the ability to shift constructively between the moods of contemplation and urgency, reflection and action. When is the time to pray, and when is the time to organize and fight? When are social arrangements minimally acceptable, and when does complicity with governing powers and principalities compromise one's own identity as a moral person? What are the proper bounds of the political and the spiritual life? Over the long history of the Church, both saints and laymen have struggled with these questions.

Finding the correct answer in any given scenario illuminates what Martha Nussbaum calls the fragility of goodness – the degree to which our ability to choose wisely and morally all too often depends on a supportive context of peers and influential institutions. Nor has the institutional church always chosen wisely. Its deeply troubling alliance with fascists in the Spanish Civil War, and its recurring failure to address the problem of child abuse, are only two examples of an all too human church failing to exemplify the moral authority of its founder. Still, the recent history of the Church does offer some beacons of moral guidance for reconciling the spiritual moods of political urgency and transcendental patience. Catholic Social Thought (CST) is a particularly rich resource here.

POLICY, PHILOSOPHY, AND THEOLOGY

Pope Francis's encyclical *Laudato Si': On Care for Our Common Home* was one of the most important intellectual and spiritual interventions in environmental policy and politics in 2015. The Pope offered a compelling moral vision for personal and ecological renewal, as well as a sweeping critique of dominant forms of consumerism (at the personal level) and cost–benefit analysis (at the level of firms and governments). The encyclical calls on all persons of good will to turn away from a selfish anthropocentrism, and toward an "integral ecology." Such an integral ecology resolutely situates human purpose and destiny in nature, refusing both dominionist doctrines of mastery over the environment, and transhumanist aspirations to transcend it. The encyclical also addresses states and policymakers directly, commending conservation and demanding a forceful response to the twin crises of anthropogenic global warming and inequality. This global vision for political economy seamlessly transitions into concluding reflections on Catholic theology and personal conversion, culminating in a "prayer for our earth" and a "Christian prayer in union with creation."

Like *Laudato Si'* itself, the essays in this volume take a dialectical approach to the concrete and the abstract, the particular and the universal, the immediate and the long term, the material and the spiritual. Neither is graspable without its opposite (or partner), just as the very concept of satisfaction or joy depends on some experience of deprivation or sadness. Theologians and philosophers have already grappled with the religious and theoretical contributions of the encyclical, and will continue do so. This volume both draws on and complements their thought, and focuses the academic conversation on *Laudato Si'* on two clusters of questions.

First, on the level of policy: how does the explicit program for conservation of land, air, and water in *Laudato Si'* fit with existing environmental programs and ideologies? Several essays in this volume compare the vision of *Laudato Si'* with those of more secular approaches to addressing environmental degradation and global poverty. They clarify what is Pope Francis's original contribution to environmental policy discourse, and what in the encyclical is drawn from extant political manifestos, scientific observations, and moral and religious ideals. These authors help us appreciate how and why the Vatican decisively "takes a side" in some contemporary struggles over labor rights and environmental degradation, while eschewing a clear position on others.

Second, at a higher level of abstraction: does *Laudato Si'* invite us to adopt a new, more capacious normative framework for addressing policy concerns? At this methodological and philosophical level, the essays illuminate a program more distant from immediate political concerns, but also more encompassing: how we frame and understand profound social and environmental problems. These are questions of methodology, inviting us write and think about climate change and poverty in ways that are not now commonly pursued either in public debates, or in elite white papers, appellate opinions, and policy journals.

These two levels of concern – about policy, and about philosophy (i.e., the types of arguments and methods that are acceptable or convincing in policy contexts) – are reflected in several of the questions that animate essays in this volume. Do papal pronouncements operate primarily at the level of philosophical justification, articulating a robust, substantive alternative to arid and abstract technocracy? Does *Laudato Si'* suggest something more than supererogatory duties for legislators and administrators – and, if so, what are they? Is the Pope's approach more like that of the US National Environmental Policy Act – requiring careful planning before humanity does more to dramatically change the planet? Or does it reflect the aims of laws like the US Endangered Species Act, imposing substantive duties to maintain certain creatures, habitats, and ecosystems? Does it demand an approach even more foundational, raising not just policy questions but fundamental issues of political economy and social justice? These questions challenge readers to consider the tone and intent of the encyclical. For those seeking guidance on critical areas of environmental policy and the economy and culture which enframe it, the essays in *Care for the World* should serve as a vital starting point toward exegesis of Pope Francis's vital work.

SOCIAL CHANGE AND AGENCY IN CATHOLIC SOCIAL THOUGHT

Every morally serious thinker has acknowledged the importance of climate change and global poverty. Sadly, though, in many elite circles, a further mark of "seriousness" is a commitment to deploy a narrow set of technology-driven and market-focused "solutions" to these issues. On this view, technology is the proximate cause of humanity's material well-being, and capitalist markets underwrite its advance. States should price carbon emissions to reflect their true cost, and international agreements should articulate a rational framework of incentives to promote greener technology. Meanwhile, more global trade in goods and services will enable workers in the developing world gradually to earn wages comparable to those of workers in developed countries. Those developed-country workers, rather than resisting global competition for jobs, should retrain themselves to learn how to provide higher-value labor. Thus, in neoliberal technocracy, both climate change and deep poverty are problems to be solved by recalibrating legal rules to optimize incentives.

Seismic political change has recently undermined both neoliberal narratives. Voters in many developed countries are challenging free trade and global institutions. As automation advances, they do not trust that education and hard work will be enough to keep them "competitive" in global markets. The rise of the gig economy and precarious labor arrangements fuels economic anxiety, as more economic gains are funneled to a narrow elite. Trade agreements are increasingly in question, as are regionalist projects like further national integration into the European Union. International cooperation on climate change appears to be unraveling, as the Trump Administration aggressively undermines the Paris Climate Change Agreement. That international tension will likely undermine national initiatives to control emissions, further exacerbating an already difficult collective action problem.

But technocratic neoliberalism is a robust ideology and, just as some of its political elements fade, its technological dimensions are advancing in scope and power. Some hope that markets for clean energy technology will continue to grow, regardless of political headwinds. Climate activists in the United States may despair at recent election results, but many point to cheapening solar and wind power as a market-driven assault on the coal industry. Some even embrace geoengineering methods to undo the effects of excess carbon emissions – including aerosolized chemicals in the atmosphere or whitening clouds to deflect sunlight.[1] Where social engineering has failed, chemical or physical engineering may succeed – but at the price of changing weather, or the chemical composition of a great deal of air and water.[2] For those versed in the potential unintended consequences of geoengineering, this approach seems about as meritorious as the development of new drugs to address the baleful consequences of geriatric polypharmacy. But for true technophiles, there is an almost religious faith in the power of "geoengineers" to ameliorate or even reverse anthropogenic global warming in a safe way.

Several essays in this volume confront this strange mixture of technocratic ambition and politico-economic fatalism. Francis's first encyclical is part of a long-standing effort to move CST beyond a *critique* of social institutions to a more positive and constructive vision of their future direction.[3] Anthony Annett engages directly with the encyclical's positive vision of the future of social institutions while retaining the perspective of spiritual subsidiarity in his chapter, "Our Common Responsibility for Our Common Home: The Activist Vision of *Laudato Si'*." Annett situates *Laudato Si'* as a continuation of the project begun in *Rerum Novarum* and *Gaudium et Spes*.[4] In those social encyclicals, the Church puts forward a framework that suggests a role for both institutional and bottom-up, personal corrective actions. For Annett, *Laudato Si'* follows from and expands on that tradition. Francis is addressing the "new things" that have emerged over the past 200 years, as global economic development and a concomitant demographic expansion push humanity against ecological boundaries, just as Pope Leo XIII analyzed the newly emerging global economic order and the urgent social questions raised regarding the relationship between capital and labor.[5] And in the same way that

Gaudium et Spes urges Catholics to read the "signs of the times," *Laudato Si'* identifies particular actions each person can take to alleviate the human and social degradation occasioned by climate change.

Annett regards Francis's discernment of the signs of the times in *Laudato Si'* as substantially in line with contemporary scientific findings about the "planetary boundaries" we are in the process of crossing as we experience Anthropocenic turbulence. *Laudato Si'* analyzes the human and environmental costs of our economic activity and "throwaway culture" in terms that are concrete and demand an immediate moral response. For example, as Annett observes, Francis devotes an entire section of the encyclical to addressing issues of water poverty, and exhibits similar concern when speaking about climate refugees and the loss of biodiversity. By focusing on the long-term costs of short-term economic decision-making, *Laudato Si'* "takes a scientific reality and endows it with moral meaning," according to Annett, underscoring the heightened need for interdependence in the face of truly global challenges.

That is a bold move for the Church, and one only taken reflectively and prayerfully, given the history of religious interventions in political struggles. Fear of the Church being itself suborned by the political struggles it attempts to comment on or influence, as well as a respect for the autonomy of conscience of voters and politicians, has encouraged a sliding scale of specificity in Church teaching: a declining willingness to mandate specific patterns of behavior or commitments as we move from the personal to the political. The thought of John Courtney Murray, so influential at the Second Vatican Council, crystallized a consensus toward lowering the religious stakes of political contestation. Catholic values can and should inform individuals' political preferences. But doctrine has been aimed primarily at the right conduct of personal life and church community. A spiritual subsidiarity (prioritizing the personal over the political) refracted and mitigated the political impact of CST (including its recommendations of our more familiar, political sense of subsidiarity).

By reading *Laudato Si'* in conversation with *Rerum Novarum* and *Gaudium et Spes*, Annett identifies potential roles for a diverse group of moral agents (ranging from supranational institutions, states, businesses, and community groups) to complement and build upon individual action. Francis urges all of these actors to participate in building forms of interdependence that will be necessary if we are to reorient our economic activities in a more responsible direction. At a supranational level, Annett suggests that international agreements like the Sustainable Development Goals (SDGs) and the Paris Agreement on Climate Change may be valuable bulwarks against environmental degradation, which states can assist in implementing domestically. He notes that rich nations have a particular obligation to act to pay down their "ecological debt" to poorer nations whose pattern of historical development and present emissions level make them less culpable for the harms associated with climate change. Annett urges the business community to support good jobs, to direct productive activity towards meeting human needs, and to create sustainable wealth that is distributed justly. He argues that religious communities should exercise leadership in efforts to support our common home and serve as models of responsible stewardship practices. Individuals and community organizations can take a variety of actions to raise awareness of environmental issues and to urge business and political leaders to pursue strategies that will support responsible stewardship, such as boycotts and divestment initiatives.

SPIRITUAL SUBSIDIARITY AND ITS LIMITS

Some critics of CST argue that religious leaders should stay out of political matters, asserting that the social is far more complex than the personal. However, even relatively straightforward moral guidance for individuals can be surprisingly complicated. Consider, for instance, the brief comment in *Laudato Si'* on air conditioning. "People may well have a growing ecological sensitivity but it has not succeeded in changing their harmful habits of consumption which, rather than decreasing, appear to be growing all the more. A simple example is the increasing use and power of air conditioning," the Pope observes, with a tantalizing indirectness (LS 55). Like the rich man in the Gospel, who slunk away, dispirited, when Jesus told him to give all he had to the poor, today's bourgeoisie might be troubled by the Pope's comments. But should they be?

There is a great deal of overuse of air conditioning in developed countries. Air conditioning also creates what G. A. Cohen calls expensive tastes and even needs: those who grow up without it are often less bothered by heat than those who have become dependent from an early age.[6] On the other hand, thousands of persons have died in heat waves over the past decade. Such extreme weather will predictably affect more vulnerable individuals in the future. A truly Catholic and communitarian response to such disasters probably would require a mass effort by families (and neighbors) to reach out to their loved ones (and those with no available family) to get them to cooling stations or hospitals.[7] But there is also a justifiable concern that such an ethic of care may overwhelm families and communities, or, failing that, that the much simpler remedy of assuring near-universal air conditioning could have saved many lives. Of course, to generate the electricity necessary for such air conditioning, countries may need to build more nuclear power plants – and if a Chernobyl-like incident occurs (or if we simply continue to struggle with the proper disposal of nuclear waste), the balance of considerations here could favor a more communitarian, rather than technological, approach. Thus a scenario analysis of the clash of ecological and humanitarian duties here may be very difficult, or inconclusive.

When it comes to right use of technology and resources, moral questions abound, and persist. Should the wealthy world reduce its carbon footprint by letting its apartments and homes grow colder in winter and warmer in summer? Was the immediate experience of post-Fukushima Tokyo, which raised thermostats in offices to make up for a temporary shutdown of nuclear power, normative? Is it hypocritical for individuals who presently enjoy air conditioning, to prescribe policies which may make it less accessible to those who do not?

Policymakers and planners will need to grapple with these problems, and many more. However, very specific questions like these may misunderstand the rhetorical approach of the encyclical. The idea is less to commend technological regression – turning back to an era with less or no air conditioning – than to caution technocrats of the wealthy world to ask themselves about what they might learn from cultures without air conditioning (or with much less of it than is common in, say, the United States). Are there ways in which an accommodation of weather – an expectation of less work, or more vacation time – may be a better "attunement" to nature, rather than an effort to master it? Can excess air conditioning deplete reservoirs of resilience that a person might naturally develop in its absence?[8]

These questions are not meant to discredit efforts to bring comforts of the most developed countries to the developing world. However, they undermine any premature certainty that simple efforts to maximize gross domestic product (GDP) will "maximize" human well-being, or whether such "maximization" is even a coherent project, given plural values and aims. Such questioning also undermines the self-image of technocrats as applying scientific calculi to maximize the costs and minimize the benefits of any particular policy intervention.

A longer-term vision of what makes a human life worthwhile is at the core of any coherent moral system. The practice of compassion, kindness, generosity, humility, and gratitude must be part of such a vision, even though wise policymakers should also be humble about the state's capacity directly to inculcate any of these patterns of thought, emotion, and conduct.

CATHOLIC SOCIAL THOUGHT, ECOMODERNISM, AND DEVELOPMENT GOALS

But how humble should policymakers be? In 1993, William J. Byron, S.J. observed that:

> Whatever the question ... [CST's] answer is usually framed in a few general principles accompanied by several general guidelines for programs consistent with the principles. For a universal teaching church, this is the way it has to be, I suppose. When it comes, however, to the future of Catholic social teaching, I cannot help but wonder whether the times might not require more precision of the Church and its teachers, if Catholic social thought is to have greater, even decisive impact.[9]

In their essays, Vincent Ialenti, Massimiliano Montini, and Francesca Volpe all consider the benefits and burdens of a more specifically Catholic response to the problems of climate change.

Ialenti takes up the implicit invitation in *Laudato Si'* to contrast Catholic environmentalism with a "technocratic paradigm." Ialenti engages with a subtle and well-articulated vision of technocracy, the Ecomodernist Manifesto, drafted by a global team of thought leaders in environmental science, economics, design, and other fields. Despite critiques of *Laudato Si'* by some of the Manifesto's authors, Ialenti sees not only conflict, but also continuity, between the ethical perspectives developed in each document. It is helpful to start with a reflection on this convergence – on a common sense of urgency and crisis in terms of climate challenges and underdevelopment – before reflecting on their diverging policy preferences and methodological approaches.

As Ialenti observes, both CST and Ecomodernism recognize climate change as a grave reality and a potentially catastrophic eventuality, deserving responses that feature "collaboration, interdisciplinarity, and big-picture thinking about the future." The two documents "provide alternate starting points for thinking about the future of the planet and humanity writ large." Both recognize that our personal and collective choices for future paths for energy development entail judgments about values. Adherents to both CST and Ecomodernism recognize extreme poverty as a scandal in a rich world – a burden that both causes suffering and vulnerability directly, and prevents human beings from reaching their full potential. And each recognizes the importance of aesthetic and spiritual responses to nature.

From this common base of concern, Ecomodernists and Francis pursue very different paths to social justice. Ialenti expertly traces these tensions, rooted in conflicting theories of history and visions of the future. For the Ecomodernists, human history is fundamentally a tale of *progress* directly attributable to *markets and technology*. Government's role is to structure law and policy so as to best incentivize technological development that simultaneously meets human needs and (secondarily) reduces human impact on nature. The core of the Ecomodernist approach is to take the present trajectory of human needs and wants as a given, and to manage the risk of various approaches to satisfying them. So, for instance, the Ecomodernist Manifesto strongly endorses the use of nuclear power, as a powerfully scalable way of meeting energy needs while decarbonizing both advanced and emerging economies.

As Ialenti observes, *Laudato Si'* has a very different perspective on both the *longue durée* of human history and the merits of accelerating technological advance. Human history seems at least as plausibly cyclical as linearly progressing. Empires rise and fall, violent periods follow peaceful ones; the great comforts of our age have shadow sides of addiction, anomie, and isolation. Whatever merits the "end of history" thesis may have once had, have now ended. Not even material technological advance is an unmixed blessing. As William Baumol observed in *The Cost*

Disease, the economic miracle of increased manufacturing productivity has a shadow side: weapons of mass destruction are easier to generate than ever.[10] This shadow side is, of course, also evident in nuclear power. Such an energy source, at present, trades the slow-but-certainly-deleterious effects of fossil fuels for a very small possibility of rapid and disastrous meltdown or toxic release. As Ialenti observes here and in other works, nuclear disposal technology has not been perfected, and the moral calculus of rapid development of such power is by no means self-evident.

Debates among Ecomodernists and adherents to CST could easily bog down into pitched battles over metaphysical commitments and the proper time frame for policy considerations. For many Ecomodernists, religious thought in the style of Pope Francis is hopelessly romantic and reactionary.[11] Meanwhile, for Catholics who dream of a "small church," or just one retreating from the sinful extravagances of modernity, it can be easy to cocoon into the satisfaction of a wisdom tradition unsullied by political compromise. Ialenti will not let either side off the hook. Instead, he gently but convincingly structures opportunities for adherents to each vision to complement and correct one another. He challenges the Ecomodernist worldview by bringing up the importance of intergenerational equity and recent disasters like Fukushima. Is Ecomodernism compatible with any version of the precautionary principle? By contrast, is the CST developed in *Laudato Si'* capable of providing guidance as to particular projects that may be undertaken in the near-term future? For example, if a less developed country announced plans to build a nuclear plant, what would be the response from those devoted to the alter-ecology articulated by *Laudato Si'*? The answer can't simply be: wait for a bit longer until better energy sources reach you. There must be some duties of developed countries to help. But what risks are acceptable to take on as such duties are fulfilled? Both Ecomodernists and Catholics can learn from one another as we struggle to answer such difficult questions.

Montini and Volpe also pursue a project of dialogue, comparing the vision in *Laudato Si'* with that embodied in the United Nations' SDGs. As they observe, the United Nations articulates seventeen broad goals for ecological protection and economic renewal, promoting 169 targets for access to clean water and renewable energy, preservation of ecosystems, and promotion of economic and social inclusion (and a time frame: they are to be met by 2030). While the Ecomodernist Manifesto's long-term vision can rightly be contrasted with that of *Laudato Si'*, most of the targets of the UN's SDGs may be framed as concrete ways of measuring whether the Vatican's concerns about matters like biodiversity, access to water, and inequality are actually being heeded by policymakers. They also provide some concrete numerical foundations for measuring our commitment to an avowed "preferential option for the poor."

Montini and Volpe situate both the SDGs and *Laudato Si'* in their respective traditions of arguments and discursive communities. They focus on the sweep and long-term perspective of *Laudato Si'* – a welcome corrective to purely economistic approaches to human and planetary well-being. As they explain, the SDGs and the recommendations in *Laudato Si'* are not merely complementary, but deeply enrich one another: "The Encyclical Letter alone . . . has not the political power of the UN 2030 Agenda to trace an enforceable sustainable development path, while the SDGs and the related UN 2030 Agenda lack the ethical afflatus and the break-through vision that permeates the Encyclical Letter."

RECOGNIZING URGENT THREATS TO HUMAN WELL-BEING

Francis's imaginative reconstruction of Mary's grieving for "the sufferings of the crucified poor and for the creatures of this world laid waste by human power" (LS 241) is not an invitation to technocratic policy analysis of the discount rate for biodiversity valuation, or how many generations of low-wage workers in Bangladesh, India, or Uganda should expect to live in squalor while the magic of the market lifts living standards. And this insistence cashes out in real recommendations for change, as in Chapter 2 of the encyclical, when the Pope approvingly quotes the bishops of Paraguay:

> Every *campesino* has a natural right to possess a reasonable allotment of land where he can establish his home, work for subsistence of his family and a secure life. This right must be guaranteed so that its exercise is not illusory but real. That means that apart from the ownership of property, rural people must have access to means of technical education, credit, insurance, and markets. (LS 94)

While *Laudato Si'* eloquently addresses what Rob Nixon calls "slow violence," it also offers important insights on more immediate threats to basic human rights. The recent rise of authoritarianism globally has endangered the lives of many vulnerable persons. For example, Philippine President Rodrigo Duterte drew international condemnation for killing at least 3,000 of his own citizens in a brutal campaign to control drug sales and trafficking. Rather than being chastised, Duterte reveled in the ensuing criticism with an incredibly offensive statement: "Hitler massacred three million Jews . . . there's three million drug addicts. I'd be happy to slaughter them." Aside from the obvious immorality of the statement, Duterte could not even get his figures right: 6 million Jews were killed in the Holocaust. His cavalier treatment of the facts reflects a depraved indifference to human suffering that is becoming all too common in today's political climate. His comrade in authoritarianism, Brazil's Jair Bolsonaro, also exemplifies the elective affinities between cruelty toward vulnerable minorities (including Brazil's indigenous and lesbian, gay, bisexual, and transgender (LGBT) community) and a destructive attitude toward vulnerable nature. Some of Bolsonaro's first actions upon taking office included an assault on extant rainforest protections, an attack on indigenous rights, and a removal of LGBT concerns from the remit of Brazil's human rights ministry. This is part of a culture of cruelty that a Franciscan charism would reject outright.

A subtler, but notable, cruelty is now influencing many governments' treatment of refugees, including climate migrants. Pope Francis has spoken forcefully and directly on this issue:

> Perhaps we can . . . uncover the hand of God in current events. In this way we will perhaps be able to recognise that God, in his wisdom, sent us, in rich Europe, the hungry to be fed, the thirsty to slake their thirst, the stranger to be welcomed, and the naked to be clothed. History then shows this: if we are a people, we will certainly welcome these as our brothers; if we are merely a group of individuals, we will be tempted only to save our own skins, but we will have no continuity.[12]

This brilliant inversion of the usual stereotype of communitarianism-as-exclusivist-nationalism exhibits a great mind at work. A strong community should be open to welcoming some manageable share of desperate persons expelled from their native lands – and to support a global political economy that both slows and reduces the impact of climate change.

This flexibility of thought and spirit is also reflected in Mark Shiffman's *The Other Seamless Garment*. Shiffman reflects on Cardinal Bernardin's "Consistent Ethic of Life" vision of spiritually informed public policy. Called a "seamless garment" approach (based on John 19:23, which described the seamless robe of Jesus at the crucifixion), Bernardin's articulation drew attention to the unity of principle underlying seemingly distinct strands of church teaching. Shiffman develops an account of *Laudato Si'* as an outline "of a comparable unity of teaching regarding the human relationship to the natural created world" that has ramifications not only for our economic life, but for how we live within the dominion of technology and how we think about our dominion over our own bodies.

Shiffman offers an incisive critique of the fragmentation of human inquiry, particularly in economistic approaches that neglect deeper foundations of human dignity and happiness in order to "focus on quantifiable objects, above all money." His essay offers, instead, a "Francis alternative," praising the way that St. Francis of Assisi offered a model of "contemplation, love and respect for the goodness, beauty, and dignity of created beings." A "personal conversion" can bring us closer to St. Francis, Shiffman argues, echoing the plea of *Laudato Si'* itself to its readers to consider the deeper resonance of their political, economic, and consumption choices.

LAUDATO SI' AS A POSITIVE VISION, AND A CRITIQUE OF TECHNOLOGY

The Francis alternative also counsels "receptivity," as Shiffman observes, "an openness to awe and wonder" (LS 11). Receptivity is one of the three key terms in

Pasquale's essay, "A Critique of Mastery and an Ethics of Attunement: From *Spe Salvi* to *Laudato Si'*." Pasquale's argument, that *Laudato Si'* presents a positive social vision of humanity's attunement to the natural world, proceeds in two steps. First, Pasquale examines both critical and constructive themes in CST. While *Laudato Si'* is faithful to both, it is an especially important advance for a constructive vision of a more just social order. Pasquale characterizes this constructive vision as an ethic of "attunement," a counsel to both states and their citizens to take concrete steps to harmonize their patterns of life with the needs of the rest of the inhabitants of "our common home." In Pasquale's framework, an ethic of attunement has three parts: a sense of rootedness in a particular community or environment; a receptivity to the beauty and wonder of the natural world; and a commitment to rectify those aspects of human endeavor that threaten to deny to future generations the gift of nature given to us.

Pasquale's proposed ethic of attunement stands in stark contrast to the religion of technology that is described and critiqued in Zachary Loeb's essay, "Towards a Bright Mountain." Loeb situates *Laudato Si'* in a tradition of critical theory that includes both Catholic voices (like Jacques Ellul) and those with a more secular orientation (like Lewis Mumford). Loeb calls the document "a work of intense, even radical, social criticism in the fine raiment of a papal encyclical," praising Pope Francis for his willingness to "expressly link environmental destruction and the neglect of the poor with the allegiance to high-tech devices." Like Francis's rhetorical bridge to an inclusive communitarianism, inverting the usual charges of exclusivism made against it, Loeb's characterization of *Laudato Si'* as a "heresy" against faith in technology directly confronts the usual deflective strategy of relegating encyclicals to work of purely sectarian interest.[13] Like any other human beings, those at the commanding heights of the most advanced technologies must take some explicit or implicit position on ultimate questions of what gives life meaning and value. The increasingly trans- and post-humanist tenor of such positions among a Silicon Valley elite accentuates Loeb's insights.

As both Loeb and Pasquale argue, tactics of indefinite life extension for humans (or, more realistically, the tiny sliver of plutocrats able to afford them) evince an attitude of mastery toward one's own body that mirrors a technocratic attitude toward the natural environment. Such an attitude includes a willingness to reject embodiment altogether, via identification with inorganic substrates supposedly capable of indefinitely maintaining "brain downloads." Countering such radical futurism, Pope Francis offers a path to cultivating ethical reflection that should give us pause both at the untrammeled development of markets, and the proliferation of technology meant to ameliorate their worst effects.

For an example of such technology, consider Harvard's "RoboBee," a microdrone that could pollinate plants in areas where bee populations have collapsed. The RoboBee may be a great pollinator – but do its programmers also intend it to mimic all the other roles bees play in ecosystems? Do we even know the full extent of such roles? And given the accelerating pace of environmental change that results from rapidification of commerce, can we be sure that we are accurately predicting the consequences of such an intervention? Whatever the merits of Hayek's "knowledge problem" as a critique of social planning, there is a wisdom to precautionary principles for large-scale ecological interventions. And we may even wish to exercise a healthy caution about our ability to judge which interventions will end up "large-scale" once unleashed upon fragile ecosystems.

Nor should aesthetic and spiritual attachments to nature be dismissed as soft-headed romanticism. On a personal level, we might experience a garden buzzing with RoboBees as a tragedy – a site of loss. Admittedly, a child born in a bee-less world would never know what he had missed. But that potential ignorance of the past – or indifference to parts of it now fading – is not a reason to consign our present to radical change without exploring all possible avenues for respecting the earth, by decelerating development now trashing our lands, waters, and skies. *Laudato Si'* implicitly endorses that deceleration, by critiquing the "rapidification" so common in contemporary life "when it causes harm to the world and to the quality of life of much of humanity" (LS 18). To be sure, there is no eternal and pristine landscape to save – the environment evolves over time, as the Ecomodernists emphasize. But that span of time was immense compared with the pace of ecological devastation apparent today. As *Laudato Si'* reminds us, nature is not simply a standing reserve to be manipulated to our ends. Its intrinsic value is distinct from its instrumental use as ecosystem services.

Alessandro Spina's essay, "*Laudato Si'* and Augmented Reality: In Search of an 'Integral Ecology' for the Digital Age," takes a closer look at the encyclical's expanded conception of the environment as encompassing networks of relationships and "social as well as natural space." In this frame, digital technologies like augmented reality (AR) should properly be regarded as forming an interconnected part of the real world, not just a parallel dimension.

Applying the insights of scholars of science and technology studies, Spina considers how the encyclical speaks about technology both as a mediator of the relationship between humans and the natural world and as a bearer of moral significance. For Spina, one of the most striking insights in the encyclical is the recognition that technology is never value-neutral, but instead embeds and propagates particular social and political values. By introducing a view of technology as both part of the environment and a site of contestation rather than something given, *Laudato Si'* underscores the importance of public deliberation and discourse about the design and architectural choices that shape digital technologies. If we consider the design choices that affect digital technology as forming part of the environment, then building an integral ecology will require each person, not just software developers, carefully to consider and debate questions about how digital worlds are and should be structured.

Spina's essay provides a model of the kind of critical engagement that will be necessary for those who are interested in promoting the intrinsic value of the natural world during a time of tumultuous technological progress. Will AR technology be used to encourage positive human action and responsibility, or will it merely reinforce neglect of environmental and social degradation? Can algorithms help us harness Big Data and make the most of the wealth of information that is now available to us, or will they instead reinscribe existing patterns of discrimination and marginalization? As Francis reminds us, "everything is interconnected," so our answers to these seemingly technical questions about the digital worlds we are building may have profound and unexpected consequences in the social and natural world. Spina hopes that answering these and other questions in a way that affirms human dignity and preserves the intrinsically valuable natural world will allow burgeoning AR technologies to complement and enhance, rather than supplant, our day to day experience.

REACHING TOWARD COOPERATION AND BUILDING CIVIC AND POLITICAL LOVE

Perhaps to a greater extent than any prior social encyclical, *Laudato Si'* speaks critically about the incentives created by the global market economy and the predictable, destructive consequences of our efforts to "maximize" short-term gain. In deploying this critique, Francis is mindful of recent history, invoking the memory of the global financial crisis as he cautions readers about the fate of our economic system:

> Saving banks at any cost, making the public pay the price, foregoing a firm commitment to reviewing and reforming the entire system, only reaffirms the absolute power of a financial system, a power which has no future and will only give rise to new crises after a slow, costly and only apparent recovery. The financial crisis of 2007–08 provided an opportunity to develop a new economy, more attentive to ethical principles, and new ways of regulating speculative financial practices and virtual wealth. But the response to the crisis did not include rethinking the outdated criteria which continue to rule the world. (LS 189)

By offering this warning, Francis invites us to contemplate possible alternatives to technocratic neoliberalism.

Nathan Schneider, Amanda Jaret, and David L. Gregory all take up Francis's call to rethink the criteria that govern our economic system and to consider how we might develop economic practices that are more compatible with our moral obligations. Their chapters advance alternatives to what Harvey Cox has called a "deified market," in order to resist excessive financialization of the economy. Schneider suggests that a renaissance of cooperative enterprise will promote an integral ecology while also restoring a social world that has been badly damaged by isolation and rapidification. Jaret and Gregory consider whether the workplace may be a fruitful site for developing a culture of care that is marked by fraternal acts of "civic and political love" in the context of dignified labor.

Schneider's essay, "'Truly, Much Can Be Done!': Cooperative Economics from the Book of Acts to Pope Francis," considers the promise of cooperative enterprise as an antidote to the throwaway culture that Francis laments in *Laudato Si'*. Schneider aims to recover a lost heritage of specifically Catholic cooperative enterprise that dates back to the Acts of the Apostles. He also reminds us that the notion of the "universal destination of goods" that Francis discusses in *Laudato Si'* has deep roots: Aquinas wrote about this concept in the course of defending private property. Schneider identifies cooperation as a thread unifying Catholic thought and practice in settings as diverse as the monastic tradition and the guild system. *Rerum Novarum* and Catholic social teaching had an explicit influence on networks of cooperatives like the Mondragon Corporation in Spain and the Antigonish network in Nova Scotia.

Schneider sees *Laudato Si'* as both a continuation of a longstanding tradition of Catholic cooperativism and also innovative in its introduction of the concept of stewardship as an aspect of cooperation. For Francis, Schneider explains, time is greater than space, and the importance of process cannot be overstated. Cooperative enterprise can facilitate the development of an "economy of honesty" that promotes participation, solidarity, and service to the common good. Equally importantly, cooperatives perform pedagogical and faith formation functions for those who work and participate in their management. To Schneider, a revival of Catholic cooperativism, which can be undertaken by individuals, businesses, and religious communities alike, is bound up with the hopeful mood of Francis's call for an integral ecology.

In their chapter, "Toward an Ethic of 'Civic and Political Love' in the Workplace," Amanda Jaret and David L. Gregory argue that Francis identifies a failure of fraternal love as one of the underlying economic causes of the linked risks of ecological and social crisis. They focus on the concept of "rapidification" introduced in *Laudato Si'* and the reversal of means and ends in the economic sphere. Francis argues that this reversal leads to a deified market, worshiped as an end in itself, without consideration of whether economic activity is directed is equitably addressing human needs.

Like Spina, Jaret and Gregory find the encyclical's expanded vision of the environment as including networks of relationships and the interconnection between the social and natural world to be illuminating. They see the encyclical's use of fraternal language and description of interconnected relations of mutuality as having crucial social implications. Labor, like technology, is both a way of mediating humanity's relationship with nature and a powerful site of social and moral contestation. Jaret and Gregory examine the special significance of labor in Catholic social teaching, emphasizing the cooperative and creative aspects of work that imbue it with dignity. They view the concept of rapidification as addressing not just technological changes that affect the workplace, but also the social changes that accompany the increased tempo of life and work. They detail some of the consequences of rapidification in the workplace, including un- and underemployment, the rise of the gig economy, and the fissuring and fracturing of traditional employment relationships.

Much as Schneider explores the possibilities of applying the lessons of the encyclical in the setting of cooperative enterprise, Jaret and Gregory's chapter offers the workplace as a possible site of resistance to rapidification and a unique forum for forging civic and political love and building an integral ecology. They posit worker associative activity as a source of dynamic, bottom-up change that can exemplify subsidiarity and solidarity. They consider several examples of how the workplace can be an important link between industrial activity that affects the natural world and the communities who are affected, and argue that collective bargaining, which is attuned to local considerations, can help develop more stakeholder-oriented business practices. They also emphasize that the workplace is a vital locus of social activity where workers can heed Francis's call for building intergenerational solidarity and overcoming the enervating, atomized individualism that constitutes another shadow side of increasing productivity. By encouraging fraternal acts of civic and political love in the workplace, workers and labor unions may be able to help build an economy that is attentive to the longer-term interests of stakeholders instead of the short-term interests of shareholders. And just as cooperative enterprise holds the promise of serving a pedagogical function, the workplace can similarly offer opportunities for participation, industrial democracy, and education.

TRANSCENDENTAL HORIZONS FOR GLOBAL POLITICS

Pope John Paul II was an inspiring leader for the tens of millions oppressed by communist rule. His papacy reflected both an intense devotion to prayer and a willingness to take to a global stage to call out the hypocrisy and brutality of the Soviet Union and the governments of its satellite regimes. This moral authority undermined the legitimacy of these autocracies.

It now appears that Pope Francis aspires to a similar position of moral leadership with respect to global neoliberalism. But Francis's task is in some ways more difficult than that of John Paul II, because the problem of anthropogenic global warming, while great and obvious in the long run, generates short-run horrors in a manner all too easily deflected by fossil fuel apologists and their supporters in government and media. Everyone knows who is responsible for the Gulag. But even the most extreme weather events are chalked up to chance by both meteorologists and those Naomi Oreskes memorably deemed "merchants of doubt."

Nonetheless, for those willing to listen to science, it is clear that anthropogenic climate change has already caused grave problems, with far worse eventualities on the horizon. On an almost monthly basis, past estimates of climate variation prove too conservative: temperatures and weather patterns are altering at a pace far faster than expected.[14] These changes are not simply linear in nature, slightly speeding up a long-anticipated process. Rather, each rise in the concentration of carbon dioxide in the atmosphere threatens to set off chaotic, unpredictable chains of self-reinforcing change. An Arctic that is far warmer than normal tends to lose reflective white sea ice faster; the resulting increase in water surface area better absorbs heat, which in turn melts more sea ice. Melted tundra may expose methane gas, which in turn will attract more heat in the atmosphere, which will melt more tundra. Scientists repeatedly discover such feedback loops, while equilibrating dynamics are far less common. As Pope Francis observes, "Doomsday predictions can no longer be met with irony or disdain. We may well be leaving to coming generations debris, desolation and filth" (LS 161) – or, worse, out-of-control feedback loops of warming that render climate as changeable as science fiction landscapes (such as the Trisolaran planets in Cixin Liu's *Three Body Problem*).

It is Pope Francis's great gift to us in *Laudato Si'* to meld the spiritual moods of urgency and contemplation in a profound and creative way. One core insight of the encyclical is that our identity as human beings and as spiritual seekers is intimately bound up with our evolutionary history as an animal endowed with reason and faith by the Creator.[15]

The *longue durée* spiritual perspective of *Laudato Si'* reflects how the great metaphors of our Christian faith tradition arise out of and affirm the critical importance of a stable environment of flora and fauna. The encyclical calls on us to express humble gratitude toward the institutions and environment that contributed so much to who we are. As Charles Taylor described in his magisterial *Sources of the Self*, whatever autonomy we can claim depends on an honest recognition of the degree to which our values have been shaped by institutions and environments outside ourselves. Appreciation of the beauty of the forest, or the sublime majesty of a mountain, is not simply one of multiple sensory responses available to us (such as a taste for chocolate or repulsion at cinnamon). Rather, in Taylor's terms, it is world-disclosive: a form of recognition of something of great value, however inchoately that aesthetic response may translate into moral judgment or political action. The urgency of *Laudato Si'* comes from the Pope's recognition that the human race now risks permanently altering nature. If such alteration does indeed occur, our access to the spiritual via the natural will be impaired.

INTRINSIC VALUE IN AN INSTRUMENTALIST AGE

How do claims of intrinsic value fare in the policy arena in an instrumentalist age? A technocratic logic has colonized environmental regulation in the United States, elevating cost–benefit analysis which discounts the future and rarely, if ever, adequately quantifies the diffuse benefits of a robust and healthy environment. In Europe, a precautionary principle tends to vitiate such utilitarianism, but is under attack from corporate lobbyists. Against such instrumentalist logics, Pope Francis has emphatically affirmed the intrinsic value of nature and the natural world (LS 33).

But then the question arises: how to translate the eloquence and insight of Francis's argument (and the larger tradition it is embedded in) into language appropriate for the public sphere? Ideally, CST should be a system of inquiry as respectable and important to political thought as, say, utilitarianism or deontological political theory. The contributors to this volume each show, either directly or indirectly, the relevance of explicitly theological perspectives to policymaking in an era of climate crisis. Given the prevalence of temperature anomalies in the past decade, there is no better time for frankly eschatological viewpoints on risk regulation, environmental law, and the broader question of humanity's relationship to technology and nature.

In his chapter, "Carbon Trading and the Morality of Markets in *Laudato Si'*," Eduardo M. Peñalver speaks directly to policymakers who are interested in infusing market-based regulatory reforms with the insights of CST. The chapter explores how policymakers might frame a cap-and-trade system that is more attentive to the concerns Francis raises about market-driven technocratic solutions. While Francis observes that pure market analyses may fall prey to replicating the harmful environmental and social effects of exclusion, he nevertheless takes as given that private property is desirable. Peñalver argues that by emphasizing the universal destination of goods, Francis seeks to expand access to private property as far as possible. When Francis quotes the bishops of Paraguay, he notes that they set forth "what the poor require to flourish *in addition to* private property." Yet Francis views markets as having two significant shortcomings: they have a tendency to commodify intrinsically valuable goods and to exclude and marginalize people. Peñalver argues that when Francis criticizes cap-and-trade systems and other market-based efforts to regulate emissions, he does so because market-based schemes may replicate these shortcomings by offering inadequate incentives to reduce emissions or by unfairly passing on costs to the poorest members of society.

Contrary to the more restrictive reading of the encyclical that some commentators have offered, Peñalver argues that *Laudato Si'* leaves open the possibility for actors to devise market-based approaches to combating climate change if they are appropriately crafted to address questions of distributive justice. Peñalver is optimistic that those interested in promoting the vision of integral ecology that Francis puts forward can devise a market-based scheme of upstream regulation of carbon emissions. He proposes a "Franciscan cap and trade policy" that would make initial allocations of entitlements to emit greenhouse gases in a manner that is sensitive to both transnational development history and intranational wealth distribution. He also envisions a more robust role for state actors, who can step in to allocate revenue from carbon permits to ensure that it indirectly or directly benefits the poorest members of society.

THE PRAGMATICS OF SPIRITUAL ENGAGEMENT

Accelerationists, who welcome rapid technological change in both social relations and the natural world, may find the encyclical's point of view quaint or mystical. But it is hard to see how they could consistently cordon off humanity itself from the type of rapid manipulation they would prescribe for the planet on which we evolved. Why not solve the problem of scarcity by genetically engineering human beings to be satisfied with the status quo? Why not promote deeply egalitarian social relations by medicating away envy or strivings for superiority?

A passive theory of social change (i.e., people stop abusing the environment when it gets very easy to not do so) is an important and compelling realist perspective on how to mitigate pollution, but it is also one that opens the door to radical experimentation on human sentiments and mental processes. It all too easily fits into the Promethean project of moral enhancement, now pursued by transhumanists. By contrast, a liberation theology perspective on dismantling structures of social sin is also a way of making right conduct easier, but it spreads the burden of change to both institutions and human beings. To guarantee a truly humane future, we will need institutions and economies that can keep pressures toward the mechanization, standardization, and diminution of human experience in check. Respect for the stability of nature (and human nature) will be indispensable.

Religious perspectives, like those developed in *Laudato Si'*, will be important contributors to such respect. Clifford Geertz's classic description of the religious perspective can double as a recognition of the unique perspectives CST can bring to a policy environment too often dominated by technocratic utilitarianism:

> The religious perspective differs from the common-sensical in that ... it moves beyond the realities of everyday life to wider ones which correct and complete them, and its defining concern is not action upon those wider realities but acceptance of them, faith in them. It differs from the scientific perspective in that it questions the realities of everyday life not out of an institutionalized scepticism which dissolves the world's givenness into a swirl of probabilistic hypotheses, but in terms of what it takes to be wider, nonhypothetical truths. Rather than detachment, its watchword is commitment; rather than analysis, encounter. And it differs from art in that instead of effecting a disengagement from the whole question of factuality, deliberately manufacturing an air of semblance and illusion, it deepens the concern with fact and seeks to create an aura of utter actuality. It is this sense of the "really real" upon which the religious perspective rests ...[16]

This mood and awareness can inform policy as constructively or insightfully as cost–benefit analysis or deontology.

As James L. Heft has observed, commenting on the work of Charles Taylor, "the dilemma of modernity in the West presents an unacceptable alternative: either a commitment to various traditional religious visions that have provided great spiritual fruit as well as untold human suffering or a commitment to an Enlightenment naturalism, or 'exclusive humanism,' which suppresses the spiritual dimension."[17] *Laudato Si'* offers a path to heal this divide, by articulating a vision of a more spiritual and humane future. It is both steeped in Catholic imagery and concerns, and grounded in universal normative concepts that should open dialogue with persons of good will of any faith tradition, or none.

THEMATIC OVERVIEW

There have been numerous commentaries on *Laudato Si'* in law reviews, journals of opinion, and theological fora.[18] Yet we view this collection as occupying a unique position in the existing literature because of the manner in which the essays engage with both the secular and sacred significance of the encyclical. We hope that these essays will provide a model for how CST may complement and enhance existing paradigms of economic, legal, technological, and ecological decision-making. We also see the essays' focus on networks of relationships as original. Nearly all of the essays in the collection interpret the encyclical's call to build an integral ecology as encouraging a view of both social and natural relationships that is grounded in mutuality. By thinking carefully about these obligations of mutuality, the essays seek to resist the tendency toward fragmentation and short-termism in our economic and policy decision-making. Instead, they offer a view of the transformative potential of acknowledging interconnection and involving as many affected stakeholders as possible in the process of deciding how to respond to environmental challenges. We also regard this recognition of mutuality as a crucial reason that the essays in this collection uniformly seek to identify supportive roles for both individuals and larger social groupings in combating the risks of environmental and social degradation.

The essays also fit into three broad thematic areas. First, echoing our initial broad categories of policy and philosophy, two clusters of essays emerge in the first two parts of the book: those primarily concerned with direct policy implications of CST, and those more focused on articulating a philosophical or methodological alternative to dominant modes of thought. We mark this division in the first two sections of the volume. In the third and final section, contributors explore the implications of both policy and philosophy for our daily lives at work (Schneider, Jaret, and Gregory), and at leisure (Spina).

The global conversation on CST should always be welcoming to experts who explore the practical application and philosophical resonance of CST in the realms of environmental policy and poverty reduction. We have found contributors who can explicate the authority and aims of *Laudato Si'* for several audiences. We hope the book reaches experts on CST and, more broadly, becomes a trusted source of insight for all those concerned about the proper role of religious thought in public life, and those who are thinking seriously about how best to tailor our economic and ecological policymaking strategies to respond to the social and environmental crises that the encyclical describes. We also hope the Catholic Church itself will see the contributors' work as the type of inquiry it will want to consult as it continues to develop its social teachings, in a spirit of dialogue. Finally, we pray, as Francis did on Earth Day of 2017: "Lord, bring healing to our lives, that we may protect the world and not prey on it, that we may sow beauty, not pollution and destruction."

Notes

1. Eduardo Porter, "To Curb Global Warming, Science Fiction May Become Fact," *New York Times*, April 4, 2017.
2. Bill McKibben, *Eaarth: Making a Life on a Tough New Planet* (New York, NY: Henry Holt and Co., 2010).
3. That movement must be a cautious one, for both historical and philosophical reasons. Historically, the Catholic hierarchy has not always aligned itself politically in honorable ways: dictators from Europe to South America enjoyed tacit or not-so-tacit acceptance of the Vatican or lower-level officials. For decades, monarchism or feudalism appeared to many leading Catholic thinkers a better social "fit" with our religious ideals than democracy, the welfare state, or socialism. More recently, some bishops' prioritization of abortion over other social issues in evaluating politicians (and even raising threats of excommunication) has troubled Catholic voters with a broader vision of social justice.
4. *Rerum Novarum* (1891); *Gaudium et Spes* (1965).
5. Two centuries may not seem like a particularly "new" time period, but for a faith about two millennia old, it is relatively recent.
6. G. A. Cohen, *If You're an Egalitarian, How Come You Are So Rich?* (Cambridge, MA: Harvard University Press, 2001).
7. See also Eric Klinenberg, *Heat Wave: A Social Autopsy of Disaster in Chicago*, 2nd edn. (Chicago, IL: University of Chicago Press, 2015), describing how close-knit communities mitigated the worst effects of a heat wave by banding together to help the vulnerable.
8. These questions recall the set of queries Romano Guardini posed, in orderto challenge excessive valuation of technological progress at the beginning of Chapter 7 of his book *The End of the Modern World* (New York, NY: Sheed & Ward, 1956).
9. William J. Byron, "The Future of Catholic Social Thought," *Catholic University Law Review*, 42 (1993), 557–70. Note, though, that the language in the Universal Declaration of Human Rights (UDHR) is even more open ended than that in many encyclicals, and some sections are clearly in tension with one another. See, e.g., Universal Declaration of Human Rights, G.A. Res. 217 (III) A, U.N. Doc. A/RES/217 (III) at 27(1) (Dec. 10, 1948) (guaranteeing both that "[e]veryone has the right freely to participate in the cultural life of the community, to enjoy the arts and to share in scientific advancement and its benefits" and that "[e]veryone has the right to the protection of the moral and material interests resulting from any scientific, literary or artistic production of which he is the author").
10. William J. Baumol, *The Cost Disease: Why Computers Get Cheaper and Health Care Doesn't* (New Haven, CT: Yale University Press, 2012).
11. Brendan T. Foht, "Two Approaches to Climate Action," *The New Atlantis*, 47 (Fall 2015): 56–66: www.thenewatlantis.com/publications/two-approaches-to-climate-action
12. "What Pope Francis Has to Say About Dostoyevsky and Guardini," *Catholic News Agency*, November 14, 2015: www.catholicnewsagency.com/news/what-pope-francis-has-to-say-about-dos toyevsky-and-guardini-76130/

13. This is not to say "mere" sectarian interest. John Milbank has made the case for "theology as itself a social science, and the queen of the sciences for the inhabitants of the *altera civitas*, on pilgrimage through this temporary world": John Milbank, *Theology and Social Theory: Beyond Secular Reason* (Oxford: Blackwell, 1990), p. 382. But Francis clearly intends *Laudato Si'* to appeal to those of any faith, and none at all.
14. Laurence C. Smith, "Greenhouse Warming: Prepare for the Worst," *The New York Review of Books*, October 13, 2016; Tim Flannery, *Atmosphere of Hope: Searching for Solutions to the Climate Crisis* (New York, NY: Atlantic Monthly Press, 2015).
15. Alasdair MacIntyre, *Dependent Rational Animals: Why Human Beings Need the Virtues* (London: Open Court, 2001).
16. Clifford Geertz, *The Interpretation of Cultures: Selected Essays* (New York, NY: Basic Books, 1973), p. 111.
17. James L. Heft (ed.), *A Catholic Modernity* (London: Oxford University Press, 1999).
18. Two recent volumes focus on the implications of the encyclical on process-relational thought and the philosophy of Alfred North Whitehead. See John Cobb, Jr. and Ignacio Castuera (eds.), *For Our Common Home: Process-Relational Responses to* Laudato Si' (Anoka, MN: Process Century Press, 2015); David Ray Griffin, *Protecting Our Common, Sacred Home: Pope Francis and Process Thought* (Anoka, MN: Process Century Press, 2016). A forthcoming essay collection, aimed primarily at an undergraduate audience, seeks to situate the theological and ecological insights of the encyclical within the broader corpus of CST. See Vincent J. Miller (ed.), *The Theological and Ecological Vision of Laudato Si': Everything Is Connected* (London: Bloomsbury Publishing, 2017). Several journal volumes have been published with an aim toward addressing particular aspects of the encyclical. See, e.g., Journal of Corporate Citizenship, 64 (2016) (journal volume evaluating the effects of the encyclical on corporate social responsibility initiatives); Interface Theology, 1 (2015), ed. John Capper (journal volume discussing theological significance of *Laudato Si'*). A number of law review articles discussing the encyclical have also been published. See, e.g., John Copeland Nagle, "Pope Francis, Environmental Anthropologist," *Regent University Law Review*, 28 (2016), 7–48 (reading *Laudato Si'* in the context of scripture and non-Catholic Christian environmental morality); Lucia A. Silecchia, "'Social Love' as a Vision for Environmental Law: *Laudato Si'* and the Rule of Law," *Liberty University Law Review*, 10 (2016), 371–98; Albert C. Lin, "Pope Francis' Encyclical on the Environment as Private Environmental Governance," *George Washington Journal of Energy & Environmental Law*, 9 (2017).

PART I

Policy Implications of *Laudato Si'*

1

Our Common Responsibility for Our Common Home: The Activist Vision of *Laudato Si'*

Anthony Annett

Laudato Si', Pope Francis's extraordinary 2015 encyclical, can be best appreciated in light of two preceding Church documents: one from 1891 and the other from 1965.

Back in 1891, Pope Leo XIII kickstarted modern Catholic social teaching with the encyclical *Rerum Novarum*. The "new things" of this document were the unprecedented social and economic ruptures brought about by the Industrial Revolution. In this encyclical, Pope Leo vociferously condemned the abuses, inequities, and injustices of the new economic arrangements. While insisting that socialism was unacceptable, he nonetheless refused to baptize the emerging capitalist mode of production. Instead, he applied ancient moral principles rooted in both scripture and tradition to the new circumstances of the day, calling for just economic relationships between rich and poor, capital and labor, the powerful and the powerless. And this set the standard for the next 125 years of Catholic social encyclicals, of which *Laudato Si'* is the latest incarnation.

The second prism through which *Laudato Si'* can be viewed dates back to 1965. In that year, the Church issued one of the key documents of the Second Vatican Council – *The Pastoral Constitution on the Church in the Modern World*, popularly known as *Gaudium et Spes*. This text marked a new chapter in the Church's engagement with the modern world, famously declaring that the Church "has always had the duty of scrutinizing the signs of the times and of interpreting them in the light of the Gospel." This meant that the Church must "recognize and understand the world in which we live, its explanations, its longings, and its often dramatic characteristics."[1] *Gaudium et Spes* promoted a bottom-up rather than a top-down form of engagement, which was a departure even from social encyclicals such as *Rerum Novarum*. It gave Catholics a mandate to act in the world in the name of the gospel, so that they could no longer default to shutting out the world by shutting their eyes in prayer.

Laudato Si' follows in the footsteps of these revolutionary documents. It is a social encyclical in the tradition of *Rerum Novarum*, in that it addresses moral questions that arise in the functioning of the modern economy – the "new things" of our day. And it follows *Gaudium et Spes* by discerning the "signs of the times" – the devastation of the natural world flowing from a flawed notion of progress – and calling for "dialogue with all people about our common home" (LS 3) to identify corrective actions at the personal and institutional levels.

This chapter describes the diagnosis of Pope Francis in *Laudato Si'*, which is aligned with the cutting-edge scientific consensus on the state of our planet. Following that, it argues that all elements of our society – international organizations, national and local governments, businesses, civil society actors, religions, and ordinary individuals – have their own role to play in identifying and implementing solutions. The main takeaway from *Laudato Si'* is that we have a common responsibility to care for our common home, and that no one gets a free pass to shirk this responsibility.

THE DIAGNOSIS OF *LAUDATO SI'*

Pope Francis opens the encyclical with a passionate appeal to care for creation. He reminds us that St. Francis of Assisi, from whom the Pope took his name, spoke of our common home as a sister and a mother. And yet, says Pope Francis, "This sister now cries out to us because of the harm we have inflicted on her by our irresponsible use and abuse of the goods with which God has endowed her. We have come to see ourselves as her lords and masters, entitled to plunder her at will" (LS 2).

This assault on creation has a context. It can be traced to the pattern and pace of economic activity since the onset of the Industrial Revolution in the eighteenth century. "Never have we so hurt and mistreated our common home as we have in the last two hundred years," says Pope Francis (LS 53).

Before the Industrial Revolution, the impact of human activity on the environment was limited and localized. Over long centuries, there was barely any economic growth. According to the economist John Maynard Keynes, the reason was the "remarkable absence of important technical improvements and . . . the failure of capital to accumulate."[2] This all changed around 1750, when wave after wave of technological progress led to wave after wave of economic expansion, down to the present day. Each wave built on preceding innovation such as textiles, steam engines, railways, steamships, electricity, automobiles, aviation, petrochemicals, and information and communications technology.

The result of these technological leaps was a huge spike in economic activity and population growth. This process started, haltingly, in England, gradually speeding up and shifting to more regions, although the greatest acceleration only took place after the Second World War. Since the dawn of the Industrial Revolution, world population has increased ninefold, to 7 billion people. The combination of more people plus more output per person has led to a mushrooming global economy: world output is now over 200 times larger than at the outset of the Industrial Revolution. This scale of change, in so short a period of time, is staggering, and shows no sign of slowing down. By the middle of the twenty-first century, global population is expected to surpass 9 billion, and the size of the global economy could easily increase threefold.[3]

It should come as no surprise that this is putting immense pressure on the planet. This is the core practical diagnosis of *Laudato Si'*. Right at the outset, it affirms the most recent scientific evidence that, because of the scale of human activity, we are bumping up against some vital planetary boundaries and disrupting some of the Earth's core cycles of biology, chemistry, and geology.[4] While the Earth's systems tend to be resilient, the current scale of change means that we can no longer rule out large, abrupt changes. Instead, we are on the verge of leaving behind the "safe operating space" of the Holocene – that remarkable 10,000-year "long summer" of climatic stability that provided the ideal conditions for human flourishing – and entering the Anthropocene, a new geological epoch in which human activity is a major influence on the Earth's systems, with unknown and uncertain consequences.

The surest way to abandon the Holocene is to blow through the planetary boundaries. The science identifies at least nine such boundaries, which should be regarded as gateways to a perilous world: climate change; ocean acidification; overuse of freshwater resources; land-use changes; interference with the nitrogen and phosphorous cycles (caused mainly by fertilizers); ozone depletion; chemical pollution; atmospheric aerosol loading (airborne pollution from burning fossil fuels); and a rapid loss of biodiversity in the context of ruptured ecosystems. Scientists tell us that some of these boundaries have already been breached – most notably climate change, biodiversity loss, and the nitrogen cycle. The others are pushing dangerously against the thresholds of safety, and they are inter-related, in the sense that violating one boundary makes it more likely that others will be violated, too.

Although Pope Francis does not use the term "planetary boundaries," his reading of the "signs of the times" is fully consistent with this cutting-edge science. He discusses pollution in remarkably blunt terms, noting that the Earth is looking like "an immense pile of filth" (LS 21). He refers to atmospheric pollution and pollution caused by fertilizers, insecticides, fungicides, herbicides, and agrotoxins. He also notes the problems caused by the acidification of the soil and water.

Turning to climate change, the most dangerous planetary boundary of all, *Laudato Si'* affirms the "solid scientific consensus" that the Earth is warming and that this is due to the concentration of greenhouse gases in the atmosphere, mainly as a result of human activity (LS 23). Pope Francis puts the blame where it belongs: on a model of development based on the intensive use of fossil fuels. He also singles out land-use changes – principally deforestation. He also points to various tipping points and feedback loops that come with climate change, including changes in the carbon cycle: a destruction of biodiversity and the loss of vital ecosystems, an increase in ocean acidification, and the potential release of methane from the melting of the ice caps.

Laudato Si' correctly regards climate change as a global problem with severe environmental, social, economic, political, and distributional implications (LS 25). Pope Francis notes that the poor live in areas most affected by climate change, and rely on livelihoods that are most vulnerable to climate change. He understands that they will feel the worst effects from ever more severe floods, droughts, and other natural disasters, yet they lack the resources and support systems to cope with such disruption. This will lead to greater patterns of migration and ever more climate refugees, and could spur greater upheaval, conflict, and war in the years ahead.

The encyclical devotes an entire section to the issue of water. Pope Francis insists that "access to safe drinkable water is a basic and universal human right, since it is essential to human survival and, as such, is a condition for the exercise of other human rights" (LS 30). Yet he notes that water demand frequently exceeds sustainable supply, and that water poverty is a major problem in places such as Africa – where people lack access to safe drinking water and are prone to droughts that hurt agricultural production. Pope Francis discusses water quality as well as quantity, noting that unsafe water leads to disease, and that underground water sources are being despoiled by mining, farming, and industrial activity. Overall, *Laudato Si'* stresses that severe water shortage is a looming reality that could affect billions of people, and that control of water supply by multinationals could become a major source of conflict.

Finally, *Laudato Si'* turns to another planetary boundary: the loss of biodiversity, and the destruction of the web of life that sustains our planet. Interestingly, the encyclical devotes more ink to this issue than to climate change. At his most passionate, Pope Francis claims that other species have inherent value in themselves, and are not merely resources to be exploited. "Because of us," he says, "thousands of species will no longer give glory to God by their very existence, nor convey their message to us. We have no such right" (LS 33).

Once again, *Laudato Si'* acknowledges the science, highlighting the important role in sustaining the planet and human well-being played by ecosystems that require biodiversity for their productivity and healthy functioning. The encyclical singles out the great tropical forests, which are at risk from rampant deforestation, as well as aquifers and glaciers. It singles out the oceans, where marine life is under threat from overfishing, ocean warming, and acidification. It also singles out the destruction of coral reefs, which are second only to the rainforests in terms of biodiversity. Ultimately, *Laudato Si'* appreciates that biodiversity is interdependence in action. As Pope Francis puts it: "Because all creatures are connected, each must be cherished with love and respect, for all of us as living creatures are dependent on one another" (LS 42). He takes a scientific reality and endows it with moral meaning.

This theme of connectedness is a recurring one in *Laudato Si'*. In particular, Pope Francis draws attention to the deep and symbiotic relationship between environmental well-being and human well-being, which lies at the core of his theological idea of integral ecology. This marks an important new departure in Catholic social teaching. At its most basic level, it suggests that when we harm the environment, we hurt the poor, and vice versa.

One instance of this is what *Laudato Si'* refers to as an "ecological debt" between rich and poor countries. This arises because the people who suffer most from the destruction of environment shoulder the least blame. When it comes to climate change, for example, the poorest 3 billion people account for a mere 6 percent of the carbon emissions that threaten our planet. In contrast, 85 percent of emissions come from the 2.5 billion people in upper- and middle-income countries.[5] However, the ecological debt is even broader than this. It also encompasses the commercial imbalances that arise from the export of raw materials from poor to rich countries. This leads to devastation of both human and natural ecology, especially when multinational companies operate in host countries in ways that they would never do at home.

Perhaps the key diagnosis of *Laudato Si'* is that a common factor lies behind the devastation of the Earth and disdain for the poor: the idea of a "throwaway culture" that devalues both people and creation. This culture reflects a mindset that leads people to take advantage of each other and treat one another and creation as mere objects to be used in pursuit of their own self-centered and self-absorbed gratification. The same mindset devastates our common home and excludes billions of people from development as a matter of course, consigning them to extreme poverty and deprivation, and treating them as an afterthought.

Because of these connections, *Laudato Si'* insists that "we are faced not with two separate crises, one environmental and the other social, but rather with one complex crisis which is both social and environmental" (LS 139). In essence, Pope Francis is asking us to hear "both the cry of the earth and the cry of the poor" (LS 49). This gives us our mandate for action.

THE DIFFERENT LEVELS OF ACTION

The last two chapters of *Laudato Si'* are devoted to solutions. The first of these, chapter 5, focuses on the institutional level, dealing with policies to achieve integral and sustainable development. Following from this, chapter 6 focuses on the personal level, dealing with what Pope Francis called "ecological conversion."

It is no coincidence that the institutional dimension comes first. Given the gravity of the socioenvironmental crisis, the urgent priority for Pope Francis is to establish a dialogue to "help us escape the spiral of self-destruction which currently engulfs us" (LS 163). The solution that he seeks would revolve around "an integrated approach to combating poverty, restoring dignity to the excluded, and at the same time protecting nature" (LS 139).

THE PRINCIPLES OF CATHOLIC SOCIAL TEACHING

Given that *Laudato Si'* is a social encyclical, it seeks dialogue grounded in the principles of Catholic social teaching, centered on the common good, which Pope Francis sees as a "central and unifying principle of social ethics" (LS 156). His definition of the common good is taken directly from *Gaudium et Spes*: "the sum of those conditions of social life which allow social groups and their individual members relatively thorough and ready access to their own fulfillment."[6] The common good is synonymous with the good arising from the shared social life, which cannot be simply disaggregated into the sum of individual goods.[7] Neither does it permit the exclusion of any person or group.

The common good is animated by the principle of solidarity, which is one of the most important principles of Catholic social teaching. Perhaps the best definition of solidarity comes from Pope John Paul II: a "firm and persevering determination to commit oneself to the common good; that is to say to the good of all and of each individual, because we are all really responsible for all."[8] For Pope Francis, solidarity is tied to the moral imperative to adopt a preferential option for the poor and seek a more equal distribution of the fruits of the Earth and human labor.[9] It "presumes the creation of a new mindset which thinks in terms of community and the priority of the life of all over the appropriation of goods by a few"[10] and it "must be lived as the decision to restore to the poor what belongs to them."[11]

Solidarity, then, can be seen a moral response to human interdependence. This means that as the world becomes more deeply interconnected, solidarity grows in importance. Without solidarity, globalization quickly degenerates into – in the memorable phrase of Pope Francis – a "globalization of indifference." Likewise, solidarity extends to future generations, which becomes ever more important in the wake of the environmental crisis.

Going even further, the principle of solidarity can also be applied to creation itself. In his speech to the United Nations a few months after the release of *Laudato Si'*, Pope Francis argued that the environment actually had rights that must be respected. This is for two reasons: first, because human beings are part of nature and inseparable from it, so that harming nature harms human beings; and, second, "because every creature, particularly a living creature, has an intrinsic value, in its existence, its life, its beauty and its interdependence with other creatures."[12] By implication, then, human beings are called upon to show solidarity with creation and with all living creatures. This is part and parcel of Pope Francis's call for a "new and universal solidarity" (LS 14).

In Catholic social teaching, solidarity is often balanced by the principle of subsidiarity. Subsidiarity is an often-misunderstood concept that basically says that decisions should be taken at the lowest level possible and the highest level necessary.[13] More formally, it insists that higher-level communities (including the state) should assist lower-level communities, but not usurp their rightful autonomy.

This means, as theologian Meghan Clark puts it, that subsidiarity is an "instrument for analyzing and gauging the appropriate locus of authority for moral decisions."[14] It is about getting the balance right.

Subsidiarity therefore respects the dignity of the human person as a "subject who is always capable of giving something to others."[15] It respects civil society and forms of communal life that are vital paths to human flourishing. It also acts as a bulwark against large, impersonal bureaucracies and corporations, on the one hand, and laissez-faire individualism on the other. In the words of Pope Francis, it "grants freedom to develop the capabilities present at every level of society, while also demanding a greater sense of responsibility for the common good from those who wield greater power" (LS 196). This responsibility includes the state, but it also includes the private sector and civil society, too. All must orient their activities to the common good.

THE SUPRANATIONAL LEVEL

When it comes to taking action to protect our common home, *Laudato Si'* conceives of different layers of responsibility in line with solidarity and subsidiarity. The first such layer is the supranational level: the level of the international community. As Pope Francis puts it, "interdependence obliges us to think of one world with a common plan" (LS 164). This is because the environmental crisis affects every person on Earth. The warming of the planet, and the poisoning of the air, water, and soil, do not respect national boundaries. The climate, says Pope Francis, "is a common good, belonging to all and meant for all" (LS 23). Ecological problems are therefore global problems requiring global solutions.

This follows a tradition in Catholic social teaching, calling for stronger decision-making authority at the global level. Back in 1963, Pope John XXIII called for "a juridical and political ordering of the world community," especially to maintain and strengthen peace between nations.[16] In the aftermath of the global financial crisis, Pope Benedict XVI extended this insight to the economic domain, calling for a "world political authority" to manage the global economy,[17] which is especially important in an era of rapid and large-scale capital mobility.

Laudato Si' extends this thinking to the sustainable development agenda, calling for enforceable international agreements to govern all aspects of the global commons, and criticizing the lack of progress on commitments made at the Rio Earth Summit in 1992 regarding greenhouse gas emissions, biodiversity, and desertification. In light of these past failures, which can be traced to a lack of political will, Pope Francis stresses the urgency of a global agreement to "progressively and quickly" replace fossil fuels with renewable sources of energy (LS 165). He also calls for a global consensus on issues such as sustainable agriculture, renewable energy, better management of forests and oceans, and access to drinking water.

It is no accident that *Laudato Si'* was issued in 2015, which was a watershed year for sustainable development. It seeks to construct a moral foundation for the main pillars of sustainable development: poverty reduction, social inclusion, and environmental sustainability. In the encyclical, Pope Francis prayed specifically for a positive outcome to negotiations, so that future generations would not "suffer the effects of our ill-advised delays." He insisted that inaction would constitute a "failure of conscience and responsibility" (LS 169).

In one sense, at least, his prayers were answered. In September 2015, the nations of the world gathered in New York to adopt the 17 Sustainable Development Goals as a guide to policy through 2030. These goals form the building blocks of global solidarity, aiming to end poverty and hunger; ensure clean water, affordable energy, healthcare and education for all; reduce inequality and make the economy work for everyone; protect the world's ecosystems and shift to a sustainable use of the Earth's resources; and build more inclusive, just, and peaceful societies (see Box 1). It was only fitting that Pope Francis addressed the United Nations on the day that these goals were adopted.

This was not the end of the story, however. Three months later, in December 2015, the nations of the world adopted the Paris Agreement on climate change. This is probably the most important international agreement since the establishment of the enduring multinational institutions in the immediate aftermath of the Second World War. Following a quarter-century of increasingly perilous inaction and acrimony in the wake of the Rio summit, world leaders finally got serious about climate change, agreeing to peak greenhouse gas emissions as soon as possible, with the goal of moving to net-zero carbon emissions in the second half of the century. This is an ambitious goal, but it is what is required to stay under the self-imposed red line of a 2-degree Celsius global warming limit.

THE DOMESTIC POLITICAL LEVEL

Although the sustainable development agenda was developed at the supranational level, its implementation must take place at the domestic national and local levels. For the agenda to succeed, the torch must be passed downwards. The Sustainable Development Goals can be achieved only when internalized into the domestic policy agenda. Under the Paris Agreement, countries are required to submit updated plans every five years, which are supposed to demonstrate a ratcheting-up of ambition. Implementation will also be assessed every five years. Thus, the national level is where the entire agenda rises or falls.

This domestic imperative applies not only to the identification and implementation of policy solutions, but also to the responsibilities that arise from global solidarity. The wealthiest countries have a stronger moral obligation to act, both because they have a stronger capacity to act and because they bear the greatest responsibility for the current state of affairs. This obligation extends to helping their

BOX 1: THE SUSTAINABLE DEVELOPMENT GOALS

Goal 1.	End poverty in all its forms everywhere.
Goal 2.	End hunger, achieve food security and improved nutrition and promote sustainable agriculture.
Goal 3.	Ensure healthy lives and promote well-being for all at all ages.
Goal 4.	Ensure inclusive and equitable quality education and promote lifelong learning opportunities for all.
Goal 5.	Achieve gender equality and empower all women and girls.
Goal 6.	Ensure availability and sustainable management of water and sanitation for all.
Goal 7.	Ensure access to affordable, reliable, sustainable and modern energy for all.
Goal 8.	Promote sustained, inclusive and sustainable economic growth, full and productive employment and decent work for all.
Goal 9.	Build resilient infrastructure, promote inclusive and sustainable industrialization and foster innovation.
Goal 10.	Reduce inequality within and among countries.
Goal 11.	Make cities and human settlements inclusive, safe, resilient and sustainable.
Goal 12.	Ensure sustainable consumption and production patterns.
Goal 13.	Take urgent action to combat climate change and its impacts.
Goal 14.	Conserve and sustainably use the oceans, seas and marine resources for sustainable development.
Goal 15.	Protect, restore and promote sustainable use of terrestrial ecosystems, sustainably manage forests, combat desertification, and halt and reverse land degradation and halt biodiversity loss.
Goal 16.	Promote peaceful and inclusive societies for sustainable development, provide access to justice for all and build effective, accountable and inclusive institutions at all levels.
Goal 17.	Strengthen the means of implementation and revitalize the Global Partnership for Sustainable Development.

Source: United Nations (2015)[18]

poorer neighbors invest in sustainable development through the transfer of technology, technical assistance, and financial resources. This should be seen less as charity and more as justice – as paying down the "ecological debt" that exists between rich and poor countries. This is especially true in the case of climate change, whereby poorer countries need the means both to shift to renewable energy and to protect themselves against the effects of climate change that can no longer be prevented.

The risk in all of this, as stressed by *Laudato Si'*, is that national interests will trump the global common good, or that a short-termist mindset will cripple economic and political action. This is because political deliberation is too often limited to the easy and the expedient – pursuing short-term economic growth at the expense of longer-term sustainable development, too timid to challenge the cult of consumerism and instant gratification, and too unwilling to cross the strong vested interests standing in the way of the energy transition. And when it comes to the Paris Agreement and the Sustainable Development Goals, the distance from these commitments – undertaken, as they are, at the global level – and the ambition embedded in them conspire to push them off the political and economic radar.

Laudato Si' does not mince words when it comes to these roadblocks. On climate change, for example, the encyclical states bluntly that "many of those who possess more resources and economic or political power seem mostly to be concerned with masking the problems or concealing their symptoms" (LS 26). "There are too many special interests," it says, "and economic interests easily end up trumping the common good and manipulating information so that their own plans will not be affected" (LS 54). Pope Francis also condemns a head-in-the-sand attitude, which serves as a license to carry on "with our present lifestyles and models of production and consumption" (LS 59).

Some commentators on *Laudato Si'* tend to play down this strong denunciation of inaction due to inadequate courage or excessive self-interest. They note that Pope Francis states explicitly that the Church "does not presume to settle scientific questions or to replace politics" (LS 188). Yet the debate and dialogue he is calling for is so that stakeholders can coalesce around real solutions that serve the common good. There is no room in this dialogue for those who sow misinformation or who actively seek to undermine sustainable development.

THE ROLE OF BUSINESS

The problems of sustainable development cannot be solved in the political realm alone. The business economy, too, has a vital role to play. As part of the dialogue needed to protect our common home, *Laudato Si'* is asking for politics and economics to unite "in the service of life." It insists that politics should not be held hostage to the economy and "nor should the economy be subject to the dictates of an efficiency-driven paradigm of technocracy" (LS 189).

Throughout the encyclical, Pope Francis is strongly critical of an economic system driven by the pursuit of profit, because he sees this mentality – with its attendant ideology based on a "magical conception of the market" (LS 190) – as a root cause of both social exclusion and environmental degradation. As far as the environment is concerned, the encyclical is certainly on solid grounds, given the well-known externality brought about by the inability of market prices to account for the social costs of economic activity. Similarly, the notion that the problems of poverty and social inclusion can be solved by market forces alone does not hold up to scrutiny.[19]

While the dominant economic paradigm insists that the main (or even the only) role of business is to maximize profits – typically equated with shareholder value – this has never been the position of Catholic social teaching. Rather, this tradition starts from the premise that corporations are communities of persons who come together to meet the needs of people. Both public and private sectors are jointly responsible for the common good, and it is not acceptable for business to focus only on the financial bottom line while outsourcing solidarity to the state.[20] What this implies is that corporations must assume a broader and wider sense of responsibility, which encompasses environmental stewardship.

Some have criticized Pope Francis for what they perceive as a certain hostility toward business. This is not the case. In fact, Pope Francis argues that business can and should be a "noble vocation" (LS 129). But to truly live out that vocation, it must serve the common good, and not just its own bottom line.

To flesh this out, the Vatican's Pontifical Council of Justice and Peace has produced a document entitled *Vocation of the Business Leader*.[21] This document points to three ways in which business can serve the common good – providing good goods, good work, and good wealth.

"Good goods" refers to the idea that business meets real human needs and supports real human flourishing – by providing "goods that are truly good and services that truly serve." It cannot be about feeding a compulsive consumerism, or what Pope Francis calls "a whirlwind of needless buying and spending" (LS 203).

"Good work" stresses the primacy of employment over profits, predicated on the idea that work is essential to human dignity and a core dimension of human flourishing. It is, as Pope Francis puts it, "part of the meaning of life on this earth, a path to growth, human development and personal fulfillment" (LS 128). On the flip side, unemployment is socially corrosive. "To stop investing in people in order to gain greater short-term financial gain," says Pope Francis, "is bad business for society" (LS 128).

Finally, good wealth means creating sustainable wealth and distributing it justly. Once again, this suggests that while profit is legitimate, it cannot be the only – or indeed the primary – goal of business. As noted, *Laudato Si'* blames this ideology of profit maximization for both an economy of exclusion and the devastation of our common home. This is a special problem when these profits arise from a failure to pay the social costs of using up shared environmental resources. Only when such costs are fully borne by those who incur them, says Pope Francis, can economic activity be considered ethical (LS 195). The implication here is that a lot of modern business, especially in the fossil fuel industry, cannot be regarded as ethical. And the best way to internalize these social costs is through properly pricing carbon, which is best accomplished by carbon taxation.

Going beyond this "negative injunction" not to pollute, business also has a "positive injunction" to support sustainable development. *Laudato Si'* notes that promoting the sustainable use of natural resources is not a waste of money, and that "more diversified and innovative forms of production which impact less on the environment can prove very profitable" (LS 191). Business can live up to its potential as a noble vacation by applying its creativity and ingenuity in pursuit of sustainable development solutions. In terms of the economy as a whole, this would be consistent with a shift away from a short-term focus on consumption and toward a longer-term focus on investment in efforts to resolve urgent problems facing the human family (LS 192). Putting it bluntly, Pope Francis is calling on business to be part of the solution, not part of the problem.

INDIVIDUALS AND COMMUNITY NETWORKS

The final chapter of *Laudato Si'* is devoted to conversion, at both the individual and communal levels. For when all is said and done, claims Pope Francis, "it is we human beings above all who need to change" (LS 202) – away from a mindset of consumerism, away from an attitude of self-centeredness and self-absorption (LS 208). In essence, the encyclical is calling on people to adopt a more balanced lifestyle and turn away from what it dubs the myths of modernity – individualism, unlimited progress, competition, consumerism, and the unregulated market (LS 210).

In this regard, *Laudato Si'* is calling for a new "ecological citizenship" infused with new ecological virtues. This goes far deeper than merely enacting and respecting laws and regulations. Rather, it centers on inculcating the intrinsic motivation to live in a certain way. After all, a virtue is simply a good habit that can be perfected only by practice. These efforts can be small, as Pope Francis notes, but they are nonetheless worthwhile – they allow people to live more fulfilled lives and they "call forth a goodness which, albeit unseen, inevitably tends to spread" (LS 212).[22] The message here is that virtue can be infectious – the good habits of one person can encourage good habits in others, which can in turn mold better social norms.

Even so, *Laudato Si'* makes clear that individual initiative alone is not enough, as uncoordinated actions cannot solve complex social problems. This ability of disconnected individuals to solve collective problems is even more limited in modern globalized economies, where power is increasingly concentrated in corporate and bureaucratic structures. So to be truly effective, individuals must band together in a coordinated communal endeavor. Ecological conversion entails not only individual conversion, but also community conversion (LS 219).

This gels well with the importance of civil society in the Catholic tradition, from the vantage point of both solidarity and subsidiarity. As regards subsidiarity, civil society occupies that vital intermediary space between the atomized individual and the impersonal state. As regards solidarity, it is seen as fertile ground for human cooperation – in the words of Pope Benedict XVI, a "most natural setting for an economy of gratuitousness and fraternity."[23] It follows that "economic forms based on solidarity, which find their natural home in civil society without being restricted to it, build up society."[24]

Civil society is, therefore, an ideal domain for habituating the virtue of solidarity. It helps to build and strengthen social capital – those vital networks of trust and reciprocity that promote dialogue and a sense of common purpose, making it easier to cooperate and solve collective problems. It helps inculcate the kinds of civic virtue needed to elevate the public interest over self-interest. And since social norms tend to be malleable, developing this sense of communal connection and civic responsibility is critical for nurturing an ecological sensibility at both individual and institutional levels.[25]

The true power of civil society comes from its ability to influence politics and economics, to counteract the forces of inertia and ideology that inhibit virtuous action. Pope Francis himself is explicit on this point, praising international agencies and civil society organizations that "draw public attention to these issues and offer critical cooperation, employing legitimate means of pressure, to ensure that each government carries out its proper and inalienable responsibility to preserve its country's environment and natural resources, without capitulating to spurious local or international interests" (LS 38).

At the political level, this would include organizing, mobilizing, advocating, voting, and lobbying. At the economic level, it could take the form of "voting with the wallet" when it comes to either purchasing or investing. This latter course of action is advocated explicitly by Pope Francis, who praises consumer movements that boycott certain products. This can, he claims, change the way businesses operate, "forcing them to consider their environmental footprint and their patterns of production" (LS 206). Some economists also argue that "voting with the wallet" is a powerful force – it nudges traditional profit-maximizing companies to act more responsibly, leading to better social outcomes without the need for government intervention.[26]

This same logic could be extended to the burgeoning movement to divest from fossil fuels. Divestment is a sound financial strategy that can help the world to make a faster and smoother transition from fossil fuels toward clean energy. This is especially urgent since staying under the 2-degree Celsius global warming limit means that 80 percent of all oil, coal, and gas reserves need to stay in the ground.[27] Hence divestment is also a moral strategy, aimed at casting a moral pall over an increasingly unethical business model, shaming the powerful interests accustomed to making enormous profits with no opprobrium. And as the American civil rights and the South African anti-apartheid movements demonstrate, financial boycotts can make a real difference.

RELIGIOUS COMMUNITIES

When it comes to civil society, churches, of course, play a leading role. Indeed, faith communities tend to be some of the greatest repositories of social capital, including in the United States.[28]

When it comes to the environment, each religion offers its own religious obligation to care for the Earth and its people. For Christians, Pope Francis is clear that ecological conversion is essential: "Living our vocation to be protectors of God's handiwork is essential to a life of virtue; it is not an optional or a secondary aspect of our Christian experience" (LS 206). But *Laudato Si'* also acknowledges that this religious injunction applies to all people of faith. This is an issue that unites all religions.

The role of religion is especially important in the sustainable development agenda, because it is, at its heart, a moral agenda, orienting us toward the common good. Without this moral foundation, scientific or technical sustainable development solutions are unlikely to spur real change. It is the moral message that resonates. This was evident in 2015, given widespread agreement that *Laudato Si'* – in tandem with statements on the moral imperative to care for the Earth issued by leading representatives of Buddhism, Hinduism, Islam, Judaism, Sikhism, and indigenous religion – had a profound impact on the Paris negotiations.

It makes sense, then, for religious believers and communities to take the lead in pushing for action to protect our common home. They can lead by example in their personal lives and also in their institutional lives – for example, by "greening" their places of worship and associated institutions. They can also lead by example in their economic and political choices.

Given the timing and significance of *Laudato Si'*, Catholics would appear to have a special duty in this regard. The good news is that across regions such as Africa and Latin America, and in countries such as the Philippines, the Catholic Church and Catholic activists are internalizing the moral message of *Laudato Si'*, standing at the vanguard of action to protect our common home. For example, at a Vatican event in 2016, Cardinal Charles Bo of Yangon, Myanmar, declared that: "Unless rich countries agree to reduce the global warming, more people will die. This to me is a criminal genocide, when the poor and the weak are exposed to violent nature created by unrestricted use of fossil fuels by rich countries."[29]

However, this is not yet the case in the United States. The US bishops' conference has been slow to take up the cause of climate change and sustainable development.[30] And many Catholic groups tend to focus narrowly on issues such as abortion, sexuality, and religious liberty – they are resistant to taking up Pope Francis's banner and framing the moral debate as part of a consistent "throwaway culture," which would include calling out the excesses and abuses that flow from free market ideology. At the same time, US Catholic leaders in politics, business, finance, civil society, and education have not adequately internalized the moral imperative contained in *Laudato Si'*. They are sometimes guided less by the all-encompassing Catholic vision of the common good and more by the individualistic ethos that finds a comfortable home in the prevailing Protestant culture – giving rise to a libertarian mentality that downplays any collective action that might second-guess market outcomes. Some Catholic educational establishments and think tanks are even funded by those libertarian and fossil fuel interests most hostile to the energy transition.

A final point: there is merit not only in unilateral action by religious believers and communities but also in multireligious collaboration in the service of protecting creation. This kind of collaboration is endorsed explicitly by Pope Francis in *Laudato Si'*, when he calls on religions to "dialogue among themselves for the sake of protecting nature, defending the poor, and building networks of respect and fraternity" (LS 201). By demonstrating the ability to work together and speak with a single voice, multireligious collaboration strengthens credibility, trust, and social cohesion. It magnifies moral authority, making it easier to influence actions and policies. And the very act of dialogue itself promotes greater understanding among religions, laying the groundwork for peace, reconciliation, and shared well-being.

CONCLUSION

Laudato Si' is one of the most significant papal texts in recent times. It follows in the footsteps of *Gaudium et Spes*, that vital document from the Second Vatican Council, and also *Rerum Novarum*, the seminal social encyclical. And its importance equals that of his predecessors.

Laudato Si' is a complex document with many layers of meaning. On one hand, it is a profound spiritual meditation on the deep relationship between human beings and the natural world, suffused with the spirituality of St. Francis of Assisi. On the other hand, it is a practical document, grounded in the current scientific reality, focused on the urgency of protecting our common home. This is a radical mandate for action that so far has not been adequately heeded. And yet it must be – the very future of our planet depends on it.

Notes

1. Pope Paul VI, *Gaudium et Spes* (1965), para. 4: www.vatican.va/archive/hist_councils/ii_vatican_council/documents/vat-ii_const_19651207_gaudium-et-spes_en.html
2. John Maynard Keynes, "Economic Possibilities for Our Grandchildren" in Essays in Persuasion (London: Macmillan, 1930).
3. Jeffrey Sachs, *The Age of Sustainable Development* (New York, NY: Columbia University Press, 2015).
4. Johan Rockström et al., "A Safe Operating Space for Humanity," *Nature*, 461 (2009), 472–5.
5. Veerabhadran Ramanathan, "The Two Worlds Approach for Mitigating Air Pollution and Climate Change" in Partha S. Dasgupta, Veerabhadran Ramanathan, and Marcelo Sánchez Sorondo (eds.),

Sustainable Humanity, Sustainable Nature: Our Responsibility (Vatican City: Pontifical Academy of Sciences, Extra Series 41, May 2–6, 2014), pp. 285–300.
6. *Gaudium et Spes*, para. 26.
7. See David Hollenbach, *The Common Good and Christian Social Ethics* (New York, NY: Cambridge University Press, 2002).
8. Pope John Paul II, *Sollicitudo Rei Socialis* (1987), 38: http://w2.vatican.va/content/john-paul-ii/en/encyclicals/documents/hf_jp-ii_enc_30121987_sollicitudo-rei-socialis.html
9. Pope Francis, speech at World Meeting of Popular Movements, July 9, 2015: http://en.radiovaticana.va/news/2015/07/10/pope_francis_speech_at_world_meeting_of_popular_movements/1157291
10. Pope Francis, *Evangelii Gaudium* (2013), para. 188: http://w2.vatican.va/content/francesco/en/apost_exhortations/documents/papa-francesco_esortazione-ap_20131124_evangelii-gaudium.html
11. Ibid., para. 189.
12. Pope Francis, address to United Nations, September 25, 2015: http://w2.vatican.va/content/francesco/en/speeches/2015/september/documents/papa-francesco_20150925_onu-visita.html
13. Meghan Clark, "Subsidiarity is a Two-Sided Coin," Catholic Moral Theology, March 12, 2012: http://catholicmoraltheology.com/subsidiarity-is-a-two-sided-coin/
14. Meghan Clark, "Seeking Solidarity for Development: Insights from Catholic Social Teaching for Implementing the UN Agenda," *Journal of Catholic Social Thought*, 13 (2016), 311–28.
15. Pope Benedict XVI, *Caritas in Veritate* (2009), para. 57: http://w2.vatican.va/content/benedict-xvi/en/encyclicals/documents/hf_ben-xvi_enc_20090629_caritas-in-veritate.html
16. Pope John XXIII, *Pacem in Terris* (1963), para. 144: http://w2.vatican.va/content/john-xxiii/en/encyclicals/documents/hf_j-xxiii_enc_11041963_pacem.html
17. *Caritas in Veritate*, para. 67.
18. United Nations, "Transforming our World: the 2030 Agenda for Sustainable Development" (Resolution adopted by the General Assembly, September 25, 2015): www.un.org/ga/search/view_doc.asp?symbol=A/RES/70/1&Lang=E
19. See Anthony Annett, "The Economic Vision of Pope Francis," in Vincent Miller (ed.), *The Theological and Ecological Vision of Laudato Si': Everything Is Connected* (New York, NY: Bloomsbury/T&T Clark, 2017).
20. *Caritas in Veritate*, para. 38.
21. Pontifical Council for Justice and Peace, *Vocation of the Business Leader: A Reflection* (Vatican City, 2012): www.pcgp.it/dati/2012-05/04-999999/Vocation%20ENG2.pdf
22. Some examples he gives include avoiding plastic and paper, reducing water consumption, separating refuse, cooking only what can be consumed, showing care for other living things, using public transport or car pooling, planting trees, and turning off lights (LS 211).
23. *Caritas in Veritate*, para. 38.
24. Ibid., para. 39.
25. John Helliwell, "Social Norms, Happiness, and the Environment: Closing the Circle," *Sustainability: Science, Practice, and Policy*, 10(1) (2014), 78–84.
26. Leonardo Becchetti, "Voting With the Wallet" (University of Rome, Tor Vergata, Working Paper No. 94, 2011).
27. Bill McKibbon, "Global Warming's Terrifying New Math," *Rolling Stone*, July 19, 2012.
28. Robert Putman, *Bowling Alone: The Collapse and Revival of American Community* (New York, NY: Simon & Schuster, 2000).
29. Charles Bo, Submission to Pontifical Academy of Sciences workshop: Laudato Si' and the Path to COP22 (Vatican City, September 28, 2016): //www.pas.va/content/accademia/en/publications/scriptavaria/laudato_si_cop22/bo.html
30. See United States Conference of Catholic Bishops, *Forming Consciences for Faithful Citizenship* (Washington DC, 2016): www.usccb.org/issues-and-action/faithful-citizenship/upload/forming-consciences-for-faithful-citizenship.pdf

2

Carbon Trading and the Morality of Markets in *Laudato Si'*

Eduardo M. Peñalver[*]

INTRODUCTION

In a brief but much noted passage of *Laudato Si'*, Pope Francis criticized so-called "cap-and-trade" approaches to reducing carbon emissions. "The strategy of buying and selling 'carbon credits,'" he said, "can lead to a new form of speculation, which would not help reduce the emission of polluting gases worldwide" (LS 171). Commentators have interpreted the passage as a categorical and moralistic rejection of market-based solutions to climate change. Read within the context of the encyclical and the broader Catholic social tradition, however, it becomes clear that the Pope's critique of "cap and trade" is simultaneously more and less all-encompassing than these initial readings allow.

The Pope's objection to market-based approaches to controlling carbon emissions is closely tied to his analysis of global economic inequality. It reflects an astute appreciation of the way in which "wealth effects" arising from inequality distort the market's ability to serve as an efficient and just means of allocating the costs of environmental protection. His critique therefore echoes earlier discussions within liberation theology of the notion of "structural sin" and reinforces calls within Catholic social thought for analysis of markets always to be considered within – and at the service of – a broader moral framework. Situating Francis's discussion within these traditions makes clear that – under the right circumstances – a cap-and-trade system of emissions regulation could be consistent with the Pope's analysis in *Laudato Si'*.

In this chapter, I will briefly describe the so-called "market-based" approaches to greenhouse gas reduction that have dominated policy discussions of climate change in recent years. I will then situate Pope Francis's objection to these sorts of policy responses, both within the broader climate debate and within the tradition of Catholic social teaching. Finally, I will propose constraints that would seem to address Pope Francis's concerns.

1 A BRIEF OVERVIEW OF MARKET-BASED CLIMATE POLICIES

Climate change presents a slow-motion environmental catastrophe that is unique in both its scale and its apparent intractability. Its causes and its effects operate on a global scale over an extended period of time. Both of these features contribute to the difficulty of addressing it. The longevity of greenhouse gases in the atmosphere and lags in their impact on climate mean that reductions in emissions do not generate meaningful consequences for the climate for decades.[1] These long lags test the far-sightedness of our existing political processes. Moreover, carbon emissions anywhere in the world will contribute to climate change across the planet, and so any action taken to combat climate change must necessarily be global in its reach. Individuals (or even individual countries) – no matter how virtuously they behave – simply cannot solve the problem on their own.

But perhaps the most daunting feature of climate change is that carbon emissions go to the very core of the modern industrial economy. For at least two centuries, our civilization has been built on the energy derived from fossil fuels.[2] Unlike, say, chlorofluorocarbons, whose impact on the ozone raised similar challenges of global scale and long-term impact and yet whose emissions were successfully curtailed by the international community,[3] it is likely that reducing carbon emissions will require fundamental changes to the organization of our systems of production and consumption. Virtually every activity that modern human beings undertake involves the creation of carbon emissions.[4] The food that we eat is grown using production methods that emit carbon;[5] it is transported to us from around the world in carbon-emitting vehicles;[6] it is packaged in materials produced through methods that emit carbon;[7] we drive to the grocery store (or ride a bus or take a subway) in vehicles that either emit carbon or rely on power generation methods that do; we then take it home to a dwelling that is heated or cooled through methods that emit carbon; and cook the food on a stove that does the same.[8] In thinking about a carbon-neutral future, it is not clear how we get from here to there without dramatic changes in virtually every corner of our lives.

Over the past few decades, a broad policy consensus has emerged around the desirability of flexible approaches to implementing greenhouse gas reductions. The complexity and ubiquity of behaviors that result in carbon emissions means that attacking the problem at the ultimate moment that the carbon is emitted into the atmosphere would necessarily be highly intrusive, inefficient and expensive.[9] The most promising and comprehensive approaches to carbon emissions reductions focus on interventions far upstream in the carbon economy. Regulating upstream – e.g., when fossil fuels are extracted from the ground or converted into a usable form – leaves to the marketplace the question of how to allocate carbon emissions to particular activities.[10]

Two tools in particular have attracted broad support among climate change policy scholars – so-called "cap-and-trade" and a carbon tax.[11] Under a cap-and-trade approach, total carbon emissions would be capped at some fixed amount.[12] Entities wishing to engage in specified carbon-emitting activities would then need to procure a permit entitling them to do so. How they procured the permit would depend on the method for initially allocating the permits under the scheme. Under some approaches, they might be able to acquire permits at a government auction; under others, the government would simply give a fixed number of permits to certain emitters.[13] Once the permits were allocated, however, the "trade" element would kick in. Entities that would rather purchase a permit and emit than cut their emissions would acquire permits from existing permit-holders – those who would rather sell their permit and reduce their emissions.[14] The ability to trade permits ensures that emitters who can more cheaply reduce their emissions will do so while those for whom it is more costly to reduce emissions will pay for the permits.[15] The result is that emissions reductions are accomplished for as low a cost as possible.

Carbon taxes are similar to "cap and trade" in their flexibility, but they do not set a fixed amount of permissible carbon emissions. Instead, they impose an across-the-board tax on carbon that makes it more expensive to emit.[16] Those for whom reducing emissions is cheaper than paying the tax will do so. Those for whom paying the tax is cheaper will continue to emit.[17] By manipulating the level of the tax, governments can increase the cost of emitting carbon to achieve emissions-reduction targets while ensuring that the burdens of reducing emissions fall on those for whom doing so is least costly.

Three features of these market-based approaches are particularly relevant to Pope Francis's evaluation of them. First, they tend to make activities that involve the emission of carbon more expensive than they would be absent the schemes.[18] The mechanisms through which this occurs vary between the two approaches. Under "cap and trade," the cost is increased by the need either to reduce carbon emissions (e.g., the cost of installing less carbon-intensive technology that is more expensive than the incumbent technology) or to spend money to purchase carbon permits. Under a carbon tax, the cost is increased by the need to install new technology or to pay the tax. The extent to which these costs can be passed on to downstream consumers in the form of higher prices depends on the context, such as the availability of less carbon-intensive substitutes for the product or activity in question.[19] The ubiquity of carbon emissions in the modern industrial economy suggests that both "cap and trade" and carbon taxes would translate into broadly increased prices for consumers for everything from food to fuel and consumer products.[20]

Second, in light of the currently unequal distribution of purchasing power, the bite of these increased prices would be felt acutely by those with the least purchasing power. There is little reason to think that products consumed by the wealthy would increase in price any more rapidly than those consumed by the poor. Rather, the safest assumption is that prices would increase across the board, reducing the purchasing power of rich and poor alike.[21] But, because of the declining marginal utility of wealth, such a decrease would be felt more immediately by the poor than by the well off.[22]

Third, and cutting in the other direction, both carbon taxes and (depending on the details) a "cap-and-trade" scheme have the potential to generate significant financial resources for the state. Because the state would have the power to determine how to use those resources, the state would also have some control over the distributive consequences of both "cap and trade" and carbon taxes.

Situating Pope Francis's Views on Markets within the Catholic Social Tradition

A Catholic Teaching on Markets before Francis

Since the advent of modern Catholic social teaching with the publication of *Rerum Novarum* in 1891, the magisterium has woven a delicate balance between affirming the importance of private property and economic freedom to human dignity, on the one hand, and, on the other, asserting the necessary subservience of property rights and markets to collective regulation based on sound moral analysis of economic transactions. Although often described as a kind of third path between socialism and capitalism, the Catholic critique of the dominant economic ideologies of the twentieth century was always marked by an asymmetry. This asymmetry derived from the Popes' repeated emphasis on the importance of the institution of private ownership (including private ownership of economically productive property) to human beings' ability to flourish.

This Catholic commitment to the importance of private property led the magisterium to condemn communism in a very categorical way. Because communists reject private ownership of the means of production, the Church argued, communist economic systems are always unjust. Pope Leo XIII set the stage for this analysis when he concluded in *Rerum Novarum* that

> the main tenet of socialism, community of goods, must be utterly rejected, since it only injures those whom it would seem meant to benefit, is directly contrary to the natural rights of mankind, and would introduce confusion and disorder into the commonweal. The first and most fundamental principle, therefore, if one would undertake to alleviate the condition of the masses, must be the inviolability of private property.[23]

In contrast, from within Catholic social teaching, the flaws of capitalism have always been viewed as somewhat more contingent. If capitalism is an economic system built around private ownership and the market economy, then it is not in its essential nature opposed to justice in the same way that communism is. Rather than condemning capitalist production in the same stark terms that he used against communists, Pope Leo XIII limited himself to admonishing business owners to exercise their authority justly:

> The following duties bind the wealthy owner and the employer: not to look upon their work people as their bondsmen, but to respect in every man his dignity as a person ennobled by Christian character. They are reminded that, according to natural reason and Christian philosophy, working for gain is creditable, not shameful, to a man, since it enables him to earn an honorable livelihood; but to misuse men as though they were things in the pursuit of gain, or to value them solely for their physical powers – that is truly shameful and inhuman.[24]

In the years since *Rerum Novarum*, Catholic social teaching has always taken pains to criticize certain recurring – perhaps even pervasive – features of capitalist economic life. The encyclicals proscribe excessively low wages,[25] long working hours,[26] suppression of workers' rights to organize,[27] consumerism,[28] and the like. But Catholic teaching has described these aspects of capitalism, where they occur as accidental, not as part of its essential moral quality. Nevertheless, these shortcomings have been sufficiently widespread that Catholic teaching was able to position itself credibly as not wholly aligned with any economic system, even as its critique of communism was always far more categorical and uncompromising.

The fall of the Soviet bloc presented a challenge to this basic orientation for Catholic social teaching as a third way between two flawed poles. It has taken some time for the Church to find its voice in a unipolar world. For example, in *Centesimus Annus*, the first social encyclical published after the fall of the Berlin Wall, John Paul II continued to treat communism as an unmitigated moral evil while describing capitalism as a system whose injustice is entirely contingent on circumstance. He put it as follows:

> [C]an it perhaps be said that, after the failure of Communism, capitalism is the victorious social system, and that capitalism should be the goal of the countries now making efforts to rebuild their economy and society? Is this the model which ought to be proposed to the countries of the Third World which are searching for the path to true economic and civil progress?
>
> The answer is obviously complex. If by "capitalism" is meant an economic system which recognizes the fundamental and positive role of business, the market, private property and the resulting responsibility for the means of production, as well as free human creativity in the economic sector, then the answer is certainly in the affirmative, even though it would perhaps be more appropriate to speak of a "business economy", "market economy" or simply "free economy."[29]

In the same encyclical, John Paul II even seemed to tack a bit to the right, perhaps in order to remain on the familiar ground between opponents on the left and the right. With communism largely defunct as a global economic force, he introduced for the first time a new opponent on the left: the "Social Assistance State":

> [E]xcesses and abuses, especially in recent years, have provoked very harsh criticisms of the Welfare State, dubbed the "Social Assistance State." Malfunctions and defects in the Social Assistance State are the result of an inadequate understanding of the tasks proper to the State. Here again the principle of subsidiarity must be respected: a community of a higher order should not interfere in the internal life of a community of a lower order, depriving the latter of its functions, but rather should support it in case of need and help to coordinate its activity with the activities of the rest of society, always with a view to the common good.
>
> By intervening directly and depriving society of its responsibility, the Social Assistance State leads to a loss of human energies and an inordinate increase of public agencies, which are dominated more by bureaucratic ways of thinking than by concern for serving their clients, and which are accompanied by an enormous increase in spending. In fact, it would appear that needs are best understood and satisfied by people who are closest to them and who act as neighbours to those in need.[30]

John Paul II did caution that the fall of communism created the danger of a doctrinaire capitalism that put markets above "the service of human freedom" and out of reach of "a strong juridical framework."[31] He warned of the "risk that a radical capitalistic ideology could spread which refuses even to consider these problems, in the a priori belief that any attempt to solve them is doomed to failure, and which blindly entrusts their solution to the free development of market forces."[32] But he presented this danger as something to look out for, not as an imminent (or existent) threat.

If we were to summarize the basic orientation of Catholic social teaching toward markets prior to the arrival of Pope Francis, it would look something like the following: private ownership and markets are not perfect, but (unlike communist economic systems) they are essentially consistent with human freedom and dignity. Markets and market actors must be subordinate to moral and juridical constraints on market activity to ensure that they actually operate in the service of the common good. Dehumanizing poverty (particularly in the developing world) and alienating consumerism (particularly in the developed world) are realities that we must collectively combat, but the solution to those problems lies within a modified market system not terribly unlike the one that prevails in Western Europe.

B Francis on the Market

Although it is still relatively early in Francis's papacy, his distinctive perspective on private ownership and markets is beginning to come into focus. His discussions of private property slant far more heavily than previous popes toward admonitions about its potential pitfalls. And his insistence on the subordination of ownership to the common good is more prominent than it was for his predecessors (LS 6, 52, 93). For instance, he says:

> Whether believers or not, we are agreed today that the earth is essentially a shared inheritance, whose fruits are meant to benefit everyone. For believers, this becomes a question of fidelity to the Creator, since God created the world for everyone. Hence every ecological approach needs to incorporate a social perspective which takes into account the fundamental rights of the poor and the underprivileged. The principle of the subordination of private property to the universal destination of goods, and thus the right of everyone to their use, is a golden rule of social conduct and "the first principle of the whole ethical and social order." (LS 93)

Gone is the defense of the necessity of private property to human well being, which had been de rigueur in prior encyclicals. It is unlikely that this absence is due to a rejection of the importance of private ownership. Instead, this new tone seems more likely to result from Francis's desire to reshape the orientation of Catholic social teaching within a contemporary context that is characterized by a consensus around the institution of private property. In the absence of a communist rival, rather than defend what most people already accept, Francis chooses to focus on spelling out what the poor require in order to flourish *in addition to* private property. Quoting the Paraguayan bishops, Francis says:

> Every campesino has a natural right to possess a reasonable allotment of land where he can establish his home, work for subsistence of his family and a secure life. This right must be guaranteed so that its exercise is not illusory but real. That means that apart from the ownership of property, rural people must have access to means of technical education, credit, insurance, and markets. (LS 94)

All of this presupposes the institution of private ownership, but that institution by itself is not enough to ensure justice for the poor. The institution must be accompanied by just distribution and by other institutions, such as an educational system, banking, insurance and fair markets in which campesinos' goods can be sold for a just price.

In addition to this assertion that property by itself is not enough, Francis adds a pointed critique of the excesses of private ownership, which contribute to the miseries of the poor:

> In many places, ... productive land is concentrated in the hands of a few owners due to "the progressive disappearance of small producers, who, as a consequence of the loss of the exploited lands, are obliged to withdraw from direct production." The most vulnerable of these become temporary labourers, and many rural workers end up moving to poverty-stricken urban areas. The expansion of these crops has the effect of destroying the complex network of ecosystems, diminishing the diversity of production and affecting regional economies, now and in the future. In various countries, we see an expansion of oligopolies for the production of cereals and other products needed for their cultivation. (LS 134)

Complementing Francis's critique of private property is a critique of markets, one that is far more trenchant and thoroughgoing than the criticisms leveled by his predecessors. Francis roots his critique in a deep skepticism about the wisdom of markets. At the foundation of that skepticism are two shortcomings that Francis perceives in market arrangements.

The first problem that he identifies is the market's tendency to commodify goods that have intrinsic value. This critique comes through particularly strongly in *Laudato Si'* when Francis – using endangered species as his example – notes the gap that opens up between intrinsic and market value when natural resources are traded. "It is not enough," he says, "to think of different species merely as potential 'resources' to be exploited, while overlooking the fact that they have value in themselves" (LS 33). If market exchange does not properly account for the full value of the resources being traded and consumed, then we cannot confidently rely on market exchanges to advance the common good, and the case for more wide-ranging state intervention in economic relations becomes stronger.

The second shortcoming is both an extension of this problem of commodification and an application of it in the human realm. If market transactions are cognizant only of market values, then those transacting in markets will tend to exclude from their consideration the interests of human beings who do not themselves have at their disposal resources that the market values. Because of this, Francis says, the market generates "an economy of exclusion and inequality" in which "masses of people find themselves excluded and marginalized: without work, without possibilities, without means of escape."[33]

Earlier Catholic critiques of capitalism have treated the market as a fundamentally just system. On this earlier view, poverty and exclusion are adventitious conditions in need of state attention, but not intrinsic features of market economies. For Francis, however, exclusion and commodification do not exist at the edges of a fundamentally just market economy. Instead, they reflect the dominance within contemporary capitalism of a "'throw away' culture" characterized by widening inequality and an "idolatry" of money and free markets:

> While the earnings of a minority are growing exponentially, so too is the gap separating the majority from the prosperity enjoyed by those happy few. This imbalance is the result of ideologies which defend the absolute autonomy of the marketplace and financial speculation. Consequently, they reject the right of states, charged with vigilance for the common good, to exercise any form of control. A new tyranny is thus born.[34]

The trope of the "throw away culture" serves for Francis as a kind of economic sibling to John Paul II's "culture of death."[35] By echoing his predecessor's language, Francis seems to signal that the modern market economy is rotten at its core in a way that calls into question the entire capitalist economic system as it actually operates today. Inverting John Paul II's critique of the "social welfare state," Francis criticizes social welfare efforts, not because of their bureaucratic inhumanity, but because they do not go nearly far enough. They merely paper over deeper structural problems within the market economy:

> [S]ociety needs to be cured of a sickness which is weakening and frustrating it, and which can only lead to new crises. Welfare projects, which meet certain urgent needs, should be considered merely temporary responses. As long as the problems of the poor are not radically resolved by rejecting the absolute autonomy of markets and financial speculation and by attacking the structural causes of inequality, no solution will be found for the world's problems or, for that matter, to any problems. Inequality is the root of social ills.[36]

To summarize, then, in Francis's version of Catholic social teaching, we see a new approach for a post-communist world. Like his predecessors, Francis takes a basic appreciation of private ownership and market transactions as a starting point. But from there he quickly shifts his attention to the shortcomings of the economy we have actually built on those twin pillars. Freed of the need to make a two-front battle against *laissez-faire* capitalism and communism, he winds up with a position that is far more skeptical of modern capitalism than the views articulated by his predecessors. It is this deep-seated skepticism of property and markets as they actually function today that guides his analysis of "cap and trade."

II FRANCIS'S CRITIQUE OF "CAP AND TRADE"

The entirety of Francis's discussion of market solutions to climate change is contained within just a few sentences in section 171 of *Laudato Si'*. He says:

> The strategy of buying and selling "carbon credits" can lead to a new form of speculation which would not help reduce the emission of polluting gases worldwide. This system seems to provide a quick and easy solution under the guise of a certain commitment to the environment, but in no way does it allow for the radical change which present circumstances require. Rather, it may simply become a ploy which permits maintaining the excessive consumption of some countries and sectors.

Within a short time of the encyclical's publication, the passage became a lightning rod. Despite its somewhat confusing use of the term "carbon credits," most readers interpreted it as a swipe at all market-based approaches to reducing carbon emissions. Economists and environmentalists united to deplore the Pope's position. Robert Stavins, the director of Harvard's Environmental Economics Program, typified this reaction when he wrote the following on the *Huffington Post*:

> [Pope Francis's] unfortunate attack on the use of the market to address climate change is out of step with the thinking and the work of informed policy analysts and policy makers around the world, who recognize that we can do more, faster, and

> better with the use of market-based policy instruments – carbon taxes and/or cap-and-trade systems.[37]

Yale economist William Nordhaus concurred. "The encyclical," Nordhaus complained in an essay in the *New York Review of Books*, "overlooks the central part that markets, particularly market-based environmental policies such as carbon pricing, must play if countries are to make substantial progress in slowing global warming."[38] He went on to lament that, in *Laudato Si'*, Francis

> missed a unique opportunity to endorse one of the two crucial elements of an effective strategy for slowing climate change. He does indeed acknowledge the soundness of the science and the reality of global warming. It is unfortunate that he does not endorse a market-based solution, particularly carbon pricing, as the only practical policy tool we have to bend down the dangerous curves of climate change and the damages they cause.[39]

A careful reading of the relevant language in the encyclical, however, makes clear that Francis is not rejecting market-based climate policy any more than he rejects markets generally. Or, more properly, it makes clear that he is critical of market-based climate policies for precisely the same reasons (and to the same degree) that he is critical of markets.

Indeed, in many ways, the ubiquity of carbon emissions in the modern system of production and consumption makes climate change the environmental problem most closely aligned with Francis's "throw-away culture." On more than one occasion, Francis identifies the looming catastrophe of climate change with the prevailing "models of production and consumption" (LS 23, 26). As with that "throwaway culture," we can only combat climate change if we embrace "changes of lifestyle, production, and consumption" (LS 23).

When he turns to specific climate policies, Francis does not say that all "carbon credit" schemes must inexorably fail to reduce emissions or encourage speculation. He says that policies crafted along these lines "can" do so by simply replicating the injustices of global markets, creating "a new form of speculation" and facilitating continued "excessive consumption." Not all "cap-and-trade" schemes reinforce existing global inequalities, but poorly crafted policies might well have that effect.

In other words, in his critique of market-based climate policies, Francis is simply observing that they may partake of the same pathologies already present in global markets and may (under certain circumstances) even exacerbate those pathologies. Because of the pervasive nature of carbon emissions – with their tight connection to consumption and economic activity of all sorts – the reliance on market mechanisms to address climate change therefore calls for extreme caution. But market-based responses to climate change might be entirely consistent with the moral framework of *Laudato Si'* if they were crafted with sufficient sensitivity to Francis's broader critique of global markets.

A great deal depends on the details of the "cap-and-trade" program. For instance, a scheme in which carbon emission permits were simply auctioned off to the highest bidder would incorporate the economic inequality of the market system as a whole.[40] The wealthy would be able to purchase permits to continue to emit (or, if it were cheaper, reduce emissions). If reducing emissions were expensive, they could simply outbid the poor for the permits, and the poor would need to either pony up scarce resources to purchase permits or – if that were not feasible – cut back on their own emissions.

III RESPONDING TO FRANCIS'S CRITIQUES: WHAT WOULD A FRANCISCAN "CAP-AND-TRADE" POLICY LOOK LIKE?

What would a market-based climate policy look like if it aimed to be responsive to Francis's concerns in *Laudato Si'*? Francis himself offers some hints in the paragraphs prior to and following his critique of carbon trading. In the paragraph before, Francis argues that rich and poor countries are not equivalently situated with respect to responsibility for climate change:

> Some strategies for lowering pollutant gas emissions call for the internationalization of environmental costs, which would risk imposing on countries with fewer resources burdensome commitments to reducing emissions comparable to those of the more industrialized countries. Imposing such measures penalizes those countries most in need of development ... In this context, there is a need for common and differentiated responsibilities. As the bishops of Bolivia have stated, "the countries which have benefited from a high degree of industrialization, at the cost of enormous emissions of greenhouse gases, have a greater responsibility for providing a solution to the problems they have caused." (LS 170)

Two paragraphs later, Francis continues this theme of global inequality.

> For poor countries, the priorities must be to eliminate extreme poverty and to promote the social development of their people. At the same time, they need to acknowledge the scandalous level of consumption in some privileged sectors of their population and to combat corruption more effectively. They are likewise bound to develop less polluting forms of energy production, but to do so they require the help of countries which have experienced great growth at the cost of the ongoing pollution of the planet. (LS 172)

In these passages, Francis is suggesting the need to take into account two intersecting axes of global inequality in crafting a just response to climate change. The first is inequality between the developing and the developed world. This inequality operates both on the level of responsibility for the problem of climate change (rooted in past inequality) and inequality in ability to absorb the costs of responding to climate change (rooted in present inequality).

As to the former, nations that experienced industrialization earlier in their history – for the most part in the global North – have contributed more to the buildup of greenhouse gases in the atmosphere than those whose industrialization has come more recently. According to one 2010 study, approximately 60 percent of the climate change impact as of today is due to the emissions of just six countries: the United States, Russia, the United Kingdom, Germany, France and Japan.[41] Compounding this inequality in historic responsibility for greenhouse gas emissions is continuing economic inequality among nations, with earlier industrializing nations enjoying significant economic advantages over those who have industrialized more recently. The Pope sees both of these inequalities as relevant to the proper design of any response to climate change.

Both considerations can be coupled with market-based climate policies. One straightforward way to address these inequalities among nations is to incorporate them into discussions of the initial allocation of entitlements to emit greenhouse gases under a "cap-and-trade" system. Poorer countries and countries that bear less responsibility for existing accumulated greenhouse gases in the atmosphere (mostly the same countries) might be given higher emissions entitlements at the outset. In a global "cap-and-trade" system, these countries could sell their emissions entitlements to wealthier countries, using the money to upgrade their own technologies or to otherwise improve the standard of living of their residents. Making the initial allocations inalienable across national boundaries would also address Francis's concerns about wealth and historical responsibility, but it would prevent the kinds of wealth transfers that a global "cap-and-trade" system would facilitate.

In addition to this global distributive question, Francis expresses a concern about the distributive impact of climate policy *within* societies. That is, any policy that increases the cost of carbon emissions across the board is likely to have a regressive effect on the poorest members of a society. This is because the poor spend a greater percentage of their income on consumption than the wealthy.[42] But the same mechanisms that address global inequality among nations can also address inequality of purchasing power within a particular society.

Allocating a nation's carbon budget to the least well off has the effect of redistributing property to the poor. Those who receive the initial allocation of carbon permits could use the entitlement themselves. Alternatively, they could limit their own emissions and sell the permit to those for whom it has more value. In either case, the permit gives them resources (a valuable entitlement to emit carbon) that they did not possess before.

Within a particular society, a "cap-and-trade" scheme might be used as an affirmative vehicle to ameliorate economic inequalities. The state might, for example, auction off carbon credits but restrict the uses to which the revenue raised from those auctions could be put. (The same could be true of revenues generated from a carbon tax.) As long as the revenues were used to improve the well-being of the poorest members of the society, the climate policy would work to alleviate the kinds of market failures to which Francis objects. The revenue might be used for public education or for social welfare programs. It might also be used to engage in direct redistribution of money, such as a guaranteed minimum income or a negative income tax.

At the broadest level, climate policies designed to both make it more expensive to emit carbon and to concentrate the impact of those increases on the wealthier members of society (either within a nation state or globally) would seem to be consistent with Francis's approach in *Laudato Si'*.

IV CONCLUSION

Francis's approach to market-based climate policies in *Laudato Si'* is of a piece with his broader critique of modern capitalism. If market-based systems are not sensitive to the distributive implications of the initial allocation of carbon emission entitlements, they will merely replicate or perhaps even reinforce the pathologies of the modern capitalist economy. But that does not mean they are categorically impermissible. It is possible to design such market-based approaches to climate policy with sensitivity to distributive justice. Doing so is entirely consistent with Francis's approach in *Laudato Si'*.

Notes

* Allan R. Tessler Dean and Professor of Law, Cornell Law School. I am grateful to Kristen Ferries and Hae Jin Kang for superb research assistance.

1. Benjamin Ewing and Douglas A. Kysar, "Prods and Pleas: Limited Government in an Era of Unlimited Harm," *Yale Law Journal*, 121 (2011), 350, 369 ("[M]any of the most devastating impacts of climate change will not happen for decades or centuries hence, even though actions taken today critically affect whether they will occur."). See also Robert N. Stavins, "A Meaningful U.S. Cap-and-Trade System to Address Climate Change," *Harvard Environmental Law Review*, 32 (2008), 293, 306 (noting that the long-term nature of the climate problem "offers significant temporal flexibility regarding emission reductions" and advocating for policies that take advantage of that temporal flexibility by "setting annual emission targets that gradually increase in stringency").
2. See Jeremey Rifkin, *The Hydrogen Economy: The Creation of the Worldwide Energy Web and the Redistribution of Power on Earth* (New York, NY: J. P. Tarcher/Putnam, 2002), p. 64 ("Remove fossil fuels from the human equation and modern industrial civilization would cease to exist.").
3. Orval E. Nangle, "Stratospheric Ozone: United States Regulation of Chlorofluorocarbons," *Boston College Environmental Affairs Law Review*, 16 (1989), 531, 536–55 (explaining how chlorofluorocarbons remain in the atmosphere despite natural atmospheric cleansing for up to 120 years, and crediting the Clean Air Act, the United Nations Environmental Program, and the Montreal Protocol as the first moves toward addressing the problem).
4. Ewing and Kysar, "Prods and Pleas," 369 (noting that "all human beings and enterprises contribute – however marginally – to the phenomenon of anthropogenic climate change" and arguing that robust tort law provides an alternative to government overreach in climate change policy).
5. Agricultural activities, such as the cultivation of crops and livestock for food, contribute almost 10 percent of all carbon emissions in the United States. See Environmental Protection Agency, "Sources of Greenhouse Gas Emissions," November 6, 2016, 11:14 AM: www.epa.gov/ghgemissions/sources-greenhouse-gas-emissions#agriculture (citing livestock, especially cattle, and rice production

as significant contributors to carbon emissions); see also Reuven S. Avi-Yonah and David M. Uhlmann, "Combating Global Climate Change: Why a Carbon Tax is a Better Response to Global Warming than Cap and Trade," *Stanford Environmental Law Journal*, 28 (2009), 3, 12 (discussing how the developing world will struggle to adapt to climate change as water and food supplies become increasingly scarce).

6. *See* Rifkin, *The Hydrogen Economy*, p. 64.
7. See Ibid.
8. See Ibid.
9. Regulations that attempt to combat carbon emissions at the source are often referred to as "command and control" regulations or "conduct instruments." See Alan Carlin, "Global Climate Change Control: Is There a Better Strategy than Reducing Greenhouse Gas Emissions?" *University of Pennsylvania Law Review*, 155 (2007), 1401, 1476. Although a "command and control" strategy is relatively inexpensive to administer, it is significantly costlier to implement than market-based alternatives. See Jonathan Baert Wiener, "Global Environmental Regulation: Instrument Choice in Legal Context," *Yale Law Journal*, 108 (1999), 677, 719 (estimating "command and control's" disadvantage in abatement cost-effectiveness at $1 trillion in the global climate context).
10. See Avi-Yonah and Uhlmann, "Combating Global Climate Change," 31 (discussing how upstream regulations (those that focus on fossil fuel production) "have the greatest ability to ensure that all sources of carbon dioxide emissions are affected, because [they focus] on carbon at the point that it enters the economy").
11. See generally Stavins, "Meaningful U.S. Cap and Trade" (describing "cap and trade" as "scientifically sound, economically rational, and politically feasible"); see also Avi-Yonah and Uhlmann, "Combating Global Climate Change" (advocating for a carbon tax policy because it is simple, cost-certain, produces revenue, and provides a signaling mechanism to producers and consumers).
12. See Stavins, "Meaningful U.S. Cap and Trade," 298 ("A cap-and-trade system limits the aggregate emissions of a group of regulated sources by creating a limited number of tradable emission allowances and requiring each firm to surrender a quantity of allowances equal to its own emissions").
13. Ibid. ("The government may initially distribute allowances for free or sell them through an auction.")
14. See Ibid., 298 ("Firms' ability to trade emission allowances creates a market in which allowances migrate toward their highest-valued use").
15. See Ibid.
16. See Avi-Yonah and Uhlmann, "Combating Global Climate Change," 32 ("A carbon tax would be imposed on all oil, coal, and natural gas production in the United States, as well as all imports").
17. Michael Waggoner, "Why and How to Carbon Tax," *Colorado Journal of International Environmental Law and Policy*, 20 (2008), 1, 6 ("Individuals can change their conduct to reduce their carbon tax liability. Although some users of certain carbon products, such as gasoline, may have little ability to reduce their carbon-based purchases, other users may be flexible even in the short run, and even more users should be flexible in the long run") (advocating for a revenue-neutral carbon tax to replace or subsidize other emissions control systems such as subsidies, command-and-control regulations, and cap and trade).
18. See Avi-Yonah and Uhlmann, "Combating Global Climate Change," 30 ("A central feature of the market-based approaches, therefore, is developing a price signal for carbon that incorporates the costs of that externality and drives the market toward finding acceptable alternatives"); see also Stavins, "Meaningful U.S. Cap and Trade," 334 ("the anticipated price increases will take place gradually over much longer periods of time than did recent spikes in energy prices").
19. See Daniel Sperling and Sonia Yeh, "Low Carbon Fuel Standards," *Issues in Science and Technology*, 25.1 (2009), 57 (describing how a carbon tax would work well in the electricity sector, in which "suppliers can choose among a wide variety of commercially available low-carbon energy sources, such as nuclear power, wind energy, natural gas, or even coal with carbon capture and sequestration," but doubting its effectiveness in the transportation sector, in which oil producers "have become almost completely dependent on petroleum . . . and can't easily or quickly find or develop low-carbon alternatives"); see also Stephen Sewalk, "A Carbon Tax with Reinvestment Is WTO Compatible," *Fordham Environmental Law Review*, 25 (2014), 338, 377 ("Another advantage is that the tax would encourage companies to continue to invest in new technologies . . . [T]he carbon tax also incentivizes

consumers to request and purchase lower emission products so the tax liability passed on to them is reduced").

20. See Tracey M. Roberts, "Mitigating the Distributional Impacts of Climate Change Policy," *Washington and Lee Law Review*, 67 (2010), 209 (discussing how either a "cap-and-trade" or carbon tax system would increase the cost of gasoline, electricity, home heating fuels, and all goods manufactured or distributed using fossil fuels).
21. See Ibid.
22. See Sanford E. Gaines and Richard A. Westin, "Introduction" in Sanford E. Gaines and Richard A. Westin (eds.), *Taxation for Environmental Protection: A Multinational Study* (New York, NY: Quorum Books, 1991), pp. 1, 11 ("The regressive effect of consumption-based taxes is well known. Because poor people spend a higher proportion of their income on consumption of goods than do the wealthy (who usually devote some of their income to savings or non-consumptive expenditures), consumption-based taxes affect the poor disproportionately").
23. Pope Leo XIII, *Rerum Novarum* (1891), para. 15: http://w2.vatican.va/content/leo-xiii/en/encyclicals/documents/hf_l-xiii_enc_15051891_rerum-novarum.html
24. Ibid., para. 20.
25. Ibid., paras. 20, 39; Pope John Paul II, *Laborem Exercens* (1981), para. 19: http://w2.vatican.va/content/john-paul-ii/en/encyclicals/documents/hf_jp-ii_enc_14091981_laborem-exercens.html
26. *Laborem Exercens*, para. 19; *Rerum Novarum*, paras. 20, 39.
27. *Laborem Exercens*, para. 20.
28. Pope John Paul II, *Centesimus Annus* (1991), paras. 36, 410: http://w2.vatican.va/content/john-paul-ii/en/encyclicals/documents/hf_jp-ii_enc_01051991_centesimus-annus.html
29. Ibid., para. 42.
30. Ibid., para. 48.
31. Ibid., para. 42.
32. Ibid.
33. Pope Francis, *Evangelii Gaudium* (2013), para. 53: http://w2.vatican.va/content/francesco/en/apost_exhortations/documents/papa-francesco_esortazione-ap_20131124_evangelii-gaudium.html
34. Ibid., para. 56.
35. Pope John Paul II, *Evangelium Vitae* (1995), para. 12.
36. *Evangelii Gaudium*, para. 202.
37. Robert Stavins, "The Papal Encyclical and Climate Change Policy," *The Huffington Post Blog*, November 2, 2015: www.huffingtonpost.com/robert-stavins/the-papal-encyclical-and_b_8456424.html
38. William D. Nordhaus, "Laudato Si': On Care for Our Common Home," *The New York Review of Books*, October 8, 2015: www.nybooks.com/articles/2015/10/08/pope-and-market/
39. Ibid.
40. Cf. Richard A. Posner, *Economic Analysis of Law*, 9th edn. (New York, NY: Aspen, 2014), §1.2, at 15 ("If that distribution is unjust, the pattern of economic activities derived from it will not have a strong claim to be regarded as just either").
41. Niklas Höhne et al., "Contributions of Individual Countries' Emissions to Climate Change and Their Uncertainty," *Climatic Change*, 106 (2011), 359, 379.
42. See Gaines and Westin, "Introduction," p. 11.

3

The Need for an "Integral Ecology" in Connection with the UN Sustainable Development Goals

Massimiliano Montini and Francesca Volpe

1 INTRODUCTION

The encyclical letter "*Laudato Si'. On Care for Our Common Home,*" issued by Pope Francis in May 2015, goes beyond a purely religious relevance, touching upon the economic, ecological, legal, social, and political spheres. Thus, pollution, climate change, water scarcity, loss of biodiversity, global inequality, the throwaway culture, and the decline in quality of human life are critically analyzed by the encyclical letter as elements of the current cultural and ecological crisis.

The present contribution will first focus on the human origin of the ecological crisis, clearly acknowledged by Pope Francis, who notes that "our immense technological development has not been accompanied by a development in human responsibility, values and conscience" (LS 105). The problem, in fact, is not technology itself; rather "it is the way that humanity has taken up technology and its development according to an undifferentiated and one-dimensional paradigm" (LS 106). This is what Pope Francis refers to as the "technocratic paradigm," an (ab)use of technology made by human beings prone to accepting and sustaining "the idea of infinite or unlimited growth, which proves so attractive to economists, financiers and experts in technology" (LS 106).

Secondly, attention will be devoted to the possible solutions and alternatives to the currently dominant mainstream growth paradigm, which are proposed by the encyclical letter in order to deal with the detrimental effects of growth on the environmental media. In Pope Francis' view, awareness of the gravity of such a crisis should push humanity into embracing new habits (LS 209). In this sense, "a new dialogue about how we are shaping the future of our planet" (LS 14) is advocated in the encyclical letter, and humanity is urged to reconsider its role within Planet Earth and its relationship with the other living species. To this effect, Pope Francis proposes the adoption of an "integral ecology" and calls for an "ecological conversion," the ecological crisis being "also a summons to profound interior conversion" (LS 217).

Finally, the chapter will establish a parallel between the encyclical letter, on the one side, and the UN SDGs and the related 2030 Agenda for sustainable development, on the other side. In particular, after highlighting some differences between the two mentioned documents, we will explore and verify the existence of positive linkages between them. In fact, these two documents, which represent two recent milestones, in the ethical and the political spheres, respectively, might play a great role in the needed transition toward a more sustainable future. Moreover, their interplay could lead to the emergence of a holistic and mutually reinforced framework of action at international level.

II THE HUMAN ORIGIN OF THE ECOLOGICAL CRISIS

"It would hardly be helpful to describe symptoms without acknowledging the human origins of the ecological crisis" (LS 101). In this way, Pope Francis opens the section on the "Human roots of the ecological crisis," proposing to "focus on the dominant technocratic paradigm and the place of human beings and of human action in the world" (LS 101). In fact, if the technological advances reached by humanity in the last two centuries are undeniable, at the same time it has to be taken into account that such an "immense technological development has not been accompanied by a development in human responsibility, values and conscience" (LS 105). What the encyclical letter questions is the "undifferentiated and one-dimensional paradigm" (LS 106) that humanity has taken up, namely the technocratic paradigm. Such a paradigm is not restricted to technology; rather, it also permeates economic and political life (LS 109) and lays the foundations for the overexploitation of natural resources. However, as explicitly acknowledged by Pope Francis, "the idea of infinite or unlimited growth, which proves so attractive to economists, financiers and experts in technology ... is based on the lie that there is an infinite supply of the earth's goods" (LS 106). This is the "big lie" of our times, conveyed by the mainstream economic model sustained by a general political consensus. Actually, the goal of potentially limitless growth seems to pertain not only to the implementation of the currently mainstream economic model. In this sense, Herman Daly, who is considered "the leading sustainability theoretician,"[1] used to say that it is the most universally accepted target.[2]

However, such a "lie" of a limitless economic growth has been unmasked by science, and especially by the branch of physics known as thermodynamics. The second law of thermodynamics, the so-called "entropy law," in fact states that the availability of energy to perform useful work for human purposes is progressively reduced every time it passes through transformations, such as those brought about by economic activities.[3] Such a finding, coupled with the limited capacity of natural sinks to absorb wastes, as well as with the limited nature of environmental resources, shows the impossibility of limitless growth on a limited planet. Indeed, according to Pope Francis, "the time has come to accept decreased growth in some parts of the world, in order to provide resources for other places to experience healthy growth" (LS 193). In our view, such an exhortation from Pope Francis can be read as an ambitious, but rather concrete, proposal to put in place an operational measure for pursuing the quest for intra-generational equity, already made several years ago by the Brundtland Report,[4] which has so far remained a dead letter.

The crisis that started in 2007–08 provided a tremendous opportunity of discontinuity with such a growth model. However, its potential was not fully exploited: only limited and partial measures have been taken and "the response to the crisis did not include rethinking the outdated criteria which continue to rule the world" (LS 189). In this sense, the encyclical letter warns that "Ecological culture cannot be reduced to a series of urgent and partial responses to the immediate problems of pollution, environmental decay and the depletion of natural resources ... To seek only a technical remedy to each environmental problem which comes up is to separate what is in reality interconnected and to mask the true and deepest problems of the global system" (LS 111). Pope Francis's stance on the issue is resolved against the widespread belief that "economics and technology will solve all environmental problems" (LS 109). Such a "faith in technology" is commonly claimed by neoclassical economics as the panacea to all environmental problems.[5] However, it should be properly recognized that even technology cannot escape the second law of thermodynamics, since it necessarily needs both low entropy flows and natural resources,[6] which, as mentioned above, are limited by definition. Therefore, the myth of limitless economic growth clashes with two main obstacles that technology, as such, cannot cope with – namely, the availability of natural resources and the capacity of the environment to respond to human-made interferences.[7] Indeed, as pointed out by Barry Commoner at the beginning of the 1970s, "pollution problems arise not out of some minor inadequacies in the new technologies, but because of their very success in accomplishing their designed aims."[8] Therefore, it can be argued that the technocratic paradigm has an inherent flaw, which cannot be fixed simply by the deployment of new and more refined technology. However, it has to be highlighted that Pope Francis's critique of the technocratic paradigm "does not mean being opposed to any technological innovations which can bring about an improvement in the quality of life" (LS 187); rather, it is a critique of the idea of dominance of humans over nature, which, through the lens of the technocratic paradigm, looks at nature as a mere "object of utility" (LS 115). Profit and immediate convenience have become the leading criteria of most economic actors, and it is not realistic to "hope that those who are obsessed with maximizing profits will stop to reflect on the environmental damage which they will leave behind for future generations" (LS 190). Therefore, Pope Francis concludes that what is needed is a redefinition of the notion of progress, as "technological and economic development which does not leave in its wake a better world and an integrally higher quality of life cannot be considered progress" (LS 194). Maybe, by advocating a renewed meaning for progress, Pope Francis wants to push away the "insomnia full of nightmares of 'being left behind' – of missing the train, or falling out of the window of a fast accelerating vehicle" that the current notion of progress evokes, according to the alert wording of Zygmunt Bauman.[9]

III THE "INTEGRAL ECOLOGY" AND THE CALL FOR AN "ECOLOGICAL CONVERSION"

It is not easy to contradict and oppose the dominance of the technocratic paradigm and its detrimental environmental effects. However, the encyclical letter avoids remaining in the framework of a gloomy environmentalism, which has proven to be effective in nothing but spreading a dismal feeling.[10] Instead, Pope Francis prefers to outline a positive, optimistic way out. According to his proposal, in order to succeed in this difficult task "there needs to be a distinctive way of looking at things, a way of thinking, policies, an educational programme, a lifestyle and a spirituality which together generate resistance to the assault of the technocratic paradigm." This is due to that fact that "otherwise" – the encyclical letter aptly warns – "even the best ecological initiatives can find themselves caught up in the same globalized logic" (LS 111). In fact, in several passages the encyclical letter affirms that a renewed relationship with nature has to go hand in hand with a "renewal of humanity itself" as "there can be no ecology without an adequate anthropology" (LS 118). In this sense, in Pope Francis's view, politics and laws will not be sufficient to stop the progressive deterioration of the environment unless we address human and social degradation (LS 123 and 48) as "our relationship with the environment can never be isolated from our relationship with others and with God" (LS 119). Pope Francis advocates a "renewal of humanity itself" (LS 118) as a precondition of the renewal of humanity's relationship with nature. To this effect, the encyclical letter proposes the adoption of an "integral ecology" (LS 137). Such an integral ecology, whose precise meaning "remains somewhat elusive,"[11] is composed by an environmental, economic, social, and cultural ecology and an ecology of daily life.[12] As noted by Sean McDonagh,[13] in the previous encyclical letters *Rerum Novarum* by Pope Leo XIII (1891) and *Populorum Progressum* by Pope Paul VI (1967), the expression used was "human ecology," while Pope Francis, proposing an "integral ecology," reveals a broader concern, which extends to all living creatures and the environment at large. Such an integral ecology can also be considered a plea for interdisciplinarity and spirituality, to be relied upon in order to overcome the widespread reductionism and the piecemeal approach that characterize human actions and to reverse the culture of dominance of human beings over nature.

However, it should be noted that Pope Francis was not the first one to use such an expression. In fact, as is pointed out in the literature, "ecological integrity" had already been employed a few years earlier by the International Theological Commission and by Cardinal Tuckson.[14] Moreover, a certain concern for ecological and environmental issues had already been shown in previous papal documents, such as Paul VI's *Populorum Progressio* of 1967 and *Octogesima Adveniens*, the *incipit* of the 14 May 1971 Apostolic Letter to Cardinal Maurice Roy, also known as "A Call to Action on the Eightieth Anniversary of *Rerum novarum*."[15] Rather, it would be more correct to say, in the words of the Columban priest and missionary Sean McDonagh, that "it is a fact of recent history that the Church has been slow to recognize the gravity of the ecological problems of the earth."[16]

In our opinion, it seems that Pope Francis wishes to address the ecological crisis as a means to trigger a response to a broader crisis, which is the crisis of humanity. In this sense, it is interesting to note that Pope Francis clearly affirms that "it is no longer enough to speak only of the integrity of ecosystems. We have to dare to speak of the integrity of human life" (LS 224). The ecological crisis, in fact, "is also a summons to profound interior conversion," identified by Pope Francis as an "ecological conversion" (LS 217).[17] However, such ecological conversion has to be experienced not only at individual level, but also at community level, so as to ensure a lasting change (LS 219). To this effect, human beings are urged to rediscover "the capacity to be happy with little" and to embrace sobriety and humility (LS 222 ff.). In sum, it may be argued that the current culture of consumerism has to be fixed and reversed.

It is interesting to note that the encyclical letter is not a standalone document. Rather, it can be placed within a relevant and consolidated economic, philosophical and legal literature. In the first place, there is an immediate link (although probably an unintentional one) with the title of the book by Guido Viale, namely "La conversione ecologica. There is no alternative."[18] Here, the expression "ecological conversion" builds on the legacy of Alexander Langer's thought and his "concrete utopia," which consists in a project of a viable and, at the same time, radically innovative alternative to the *status quo*, entailing both a personal and a collective conversion.[19] In addition, a first substantial connection can be found with the work of Jean Baptiste De Foucauld and his "frugal abundance" – an expression used to designate a hoped-for society based on justice and creativity.[20] Actually, such an expression had been used for the first time several years before, in the framework of the collective work of the club "Echange e Projets" founded by Jacques Delors in 1973, and later used also by Serge Latouche, who entitled "frugal abundance" a series of essays written to clarify his degrowth programme.[21] A further connection can be established also with Herman Daly's work, in that for a long time he has been proposing to replace the neoclassical mantra "more is better" with "enough is best,"[22] in order to promote the transition to a steady state economy, characterized by sufficiency, ecological sustainability, and durability of products.[23] However, as André Gorz has correctly warned, it should be taken into consideration that a break with the tendency to produce more and consume more, as well as the proposal of a lifestyle oriented to do better with less, requires a discontinuity with "a society where we produce nothing of what we consume and we consume nothing of what we produce."[24] A profound change of attitude and of corresponding behaviors is therefore needed. However, commitment and responsibility should not be always understood as burdens; in Pope Francis's words: in fact, "such sobriety, when lived freely and consciously, is liberating" (LS 223).

The encyclical letter also addresses another very controversial and serious issue, related to the relationship between Christianity and nature. By putting ecology at the center of the stage, Pope Francis dares to renovate the traditional interpretation of Christianity as an anthropocentric religion, which has contributed over the centuries to legitimizing a predatory attitude and behavior toward nature. As is explicitly recognized by Pope Francis, "some committed and prayerful Christians, with the excuse of realism and pragmatism, tend to ridicule expressions of concern for the environment. Others are passive; they choose not to change their habits and thus become inconsistent"; therefore "what they all need is an 'ecological conversion'" (LS 217). The encyclical letter takes also a clear stance against "an inadequate presentation of Christian anthropology," which contributed to "a wrong understanding of the relationship between human beings and the world" (LS 116). According to Pope Francis, the wrong idea of "dominion" over nature should be replaced by a "sense of responsible stewardship" (LS 116). This proposal seems to echo the reasoning contained in the American Baptist Policy Statement on Ecology – An Ecological Situational Analysis, which was issued in the late 1980s, well before the encyclical letter.[25] Its section "Creation and the Covenant of Caring," in fact, after tracing the etymology of the word "stewardship" (which, coming from the Greek words for "house" and "management," is "the root word for economics and ecology"), affirms that our responsibility as stewards "implies a great degree of caring for God's creation and all God's creatures" and proposes an understanding of Christianity in which "eco-justice is the vision of the garden in Genesis – the realm and the reality of right relationship."[26]

The advocated new relationship between humanity and nature is grounded on the spirituality of Saint Francis of Assisi, which emerges throughout the encyclical letter (LS 1, 10, 11, 12, 66, 87, 91, 125, 218, 221). In fact, Saint Francis of Assisi is considered by Pope Francis "the example par excellence ... of an integral ecology lived out joyfully and authentically" (LS 10). In particular, through his life he has witnessed the inseparable connection between the planet and the people, being called "to care for all that exists" far beyond any intellectual appreciation or economic calculus (LS 11). Thus, Pope Francis shapes his ecological conversion on the legacy of Saint Francis of Assisi. In our view, this also echoes the reasoning of the contemporary environmental philosopher Mark Sagoff, who questions the idea that "every environmental problem should be understood as an economic one," holding that environmentalism should deal with social and moral values rather than computing costs and benefits.[27]

IV THE INTERPLAY BETWEEN THE ENCYCLICAL LETTER AND THE UN SUSTAINABLE DEVELOPMENT GOALS

The encyclical letter was issued in a crucial moment for the future of environmental protection and the promotion of sustainable development. Only a few months after its publication, in fact, another document, potentially of the utmost importance for the creation of a more sustainable society, was adopted by the United Nations (UN) General Assembly: the United Nations 2030 Agenda for sustainable development and the related Sustainable Development Goals (SDGs).[28] The UN 2030 Agenda is a plan of action for people, planet, and prosperity, which also aims to strengthen universal peace and promote a renewed global partnership for sustainable development. It encompasses 17 goals and 169 targets, which will be monitored through a complex set of indicators, which are still under development.[29]

The two mentioned documents, although coeval and, generally speaking, oriented to promote more sustainable development patterns, are profoundly different. In particular, as was interestingly analyzed by Klaus Bosselmann, drawing from the work of Hickel, Kirk, and Brewer, three main differences can be detected.[30] First, the approaches differ greatly: the encyclical letter is holistic, making explicit the systemic nature of the covered issues, while the UN 2030 Agenda, along with its many goals and targets, is thorough and characterized by a piecemeal approach. Secondly, throughout the encyclical letter, the idea of limitless economic growth as the main aim of society is strongly opposed and criticized; the UN 2030 Agenda, on the other hand, encompasses economic growth with a dedicated goal, namely Goal No. 8. Thirdly, the encyclical letter is a courageous document, which dares to criticize explicitly the irresponsible behavior of transnational corporations, as well as the overwhelming financial sector and, more generally, the unsustainability of the mainstream development model; such clear wording, instead, is completely missing in the UN 2030 Agenda and the related SDGs, which use much more neutral language and are still embedded in the traditional neoclassical economic way of thinking.

Despite such undeniable differences, which pose the encyclical letter well beyond the UN 2030 Agenda, in our opinion it is worth trying to verify and explore whether some positive linkages between the two mentioned documents may exist. Their interplay, in fact, could prove useful to boost the transition toward a more sustainable future, leading to the emergence of a mutually reinforced framework of action at international level.

A first similarity can be found in the scope: both the encyclical letter and the UN 2030 Agenda target the global community at large, without any distinction between rich and poor countries, differently from what happened in the previous Millennium Development Goals.[31] As a consequence, it can be argued that they are "universal" in scope in so far they call the whole of humanity to action.

Secondly, the two documents can be seen as a sign of the times. The encyclical letter, in fact, has made a tremendous breakthrough by clearly addressing the controversial issue of the relationship between Christianity and nature, explicitly recognizing the need for a radical change to overcome the traditional Christian idea of dominion of human beings over nature. Similarly, the UN 2030 Agenda and the related SDGs have been adopted in the framework of a "historic UN summit,"[32] at the end of a complex negotiating process, starting within the 2012 Rio+20 Conference, and were welcomed as a milestone in the process of strengthening universal peace with their determination "to take the bold and transformative steps which are urgently needed to shift the world onto a sustainable and resilient path."[33]

Thirdly, both documents endorse the principle of common but differentiated responsibility. In this respect, the UN 2030 Agenda seems to outdo the encyclical letter, in so far it restates Principle 7 of the Rio Declaration, which applies the principle of common but differentiated responsibilities to environmental degradation at large, while the wording (although maybe not the aim) of the encyclical letter limits the application of the principle of common but differentiated responsibilities to the climate change issue (LS 52, 170). Quite to the contrary, another fundamental principle of environmental law, namely the precautionary principle, seems to be substantially neglected by both documents. The encyclical letter, in fact, limits itself to a brief mention of such a principle when dealing with the 1992 Rio Declaration (LS 186), whereas in the UN 2030 Agenda any explicit reference to the precautionary principle is completely missing.

Finally, although Pope Francis deems politics and laws insufficient to stop the pace of environmental deterioration, as already mentioned above, the encyclical letter calls for the "establishment of a legal framework which can set clear boundaries and ensure the protection of ecosystems," by arguing that "otherwise, the new power structures based on the techno-economic paradigm may overwhelm not only our politics but also freedom and justice" (LS 53). Politics has a key role to play, being urged by Pope Francis to become "far-sighted and capable of a new, integral and interdisciplinary approach to handling the different aspects of the crisis" (LS 197). In particular, politics is called to take up the challenge of "rethinking processes in their entirety" and going beyond "few superficial ecological considerations," which would fail to question the logic underlying the current culture (LS 197). The essential role that law, regulation, and policy can play in the effective promotion of a sustainable development path[34] is also stressed throughout the UN 2030 Agenda, which contains several references to international legal documents; indeed, as affirmed in the keynote address at the Asia-Pacific Roundtable of May 2015, many of the SDGs "will hinge on improving the rule of law, especially in the area of international and national environmental law."[35]

The similarities pointed out seem to be sufficient to justify a plea for a combined reading of the two analyzed documents. In this sense, we suggest going beyond their independent legacy, by rather looking for their systemic integration, which could reinforce their communal message for the promotion of a more sustainable development model. Moreover, an integrated and merged understanding of the two documents could also help to fill their respective gaps. In fact, the encyclical letter alone has not the political power of the UN 2030 Agenda to trace an enforceable sustainable development path,[36] while the SDGs and the related UN 2030 Agenda lack the ethical afflatus and the breakthrough vision that permeate the encyclical letter. Therefore, the two documents could draw advantages from each other and, more importantly, sustainability could benefit tremendously from their combined implementation.

V CONCLUSION

Many are the critiques that could be moved to the encyclical letter from an (ecological) sustainability perspective: the absence of any reference to the controversial issue of population control, the failure sufficiently to consider the findings of the evolutionary theory, and a rather "romantic" vision of ecosystem harmony, just to mention a few.[37] However, in our opinion, in the evaluation of the encyclical letter priority should be given to the overall vision emerging from it. In fact, from a substantial point of view, the encyclical letter renovates the traditional Christian approach to the ecological question, by giving a new impetus to the promotion of more sustainable development patterns. In this sense, not only does it pose itself in discontinuity with the traditional religious approach, but rather it advocates a clear shift from the "limitless growth development model" toward "a new economy, more attentive to ethical principles" (LS 189). As seen above, such a departure from the mainstream reference scenario is proposed through the replacement of the throwaway culture and the technocratic paradigm with a profoundly renewed and "joyfully sober attitude," shaped by the integral ecology and the ecological conversion. By so doing, the encyclical letter also seems to meet the methodological requirements posed by the complexity of sustainability challenge, namely the need for a transdisciplinary vision and for integrated solutions. In this sense, by proposing a new development model, what Pope Francis advocates could be seen as a true paradigm shift. In fact, he goes beyond utilitarianism and "tyrannical anthropocentrism unconcerned for other creatures" (LS 68), by recognizing to different species and ecosystems an "intrinsic value independent of their usefulness" (LS 140, 33). In addition, he envisions a new alliance among justice, equity, solidarity, and ecological sustainability. In this sense, as affirmed in the literature, Pope Francis proposes that the encyclical letter "be taken as a starting point for dialogue about setting alternative goals for politics, working in support of substantive peace, social justice, and ecological health."[38]

In our opinion, the new paradigm outlined in the encyclical letter could be reinforced by being considered not in isolation, but rather in combination with other relevant documents on the sustainability issue. In particular, as proposed in this chapter, the encyclical letter could find an operational tool in the UN 2030 Agenda for sustainable development and in a combined reading and understanding of the two documents. In fact, in spite of the above mentioned differences, the encyclical letter and the UN 2030 Agenda for sustainable development and the related SDGs are likely to play a key role in shaping the near future, by dictating the agenda for the Christian Church and for the international community at large.

More precisely, the groundbreaking vision that permeates the programmatic text of the encyclical letter should inspire the drafting and implementation of the measures needed to attain the SDGs. This could help to fill the SDGs shortcomings, with particular reference to their lack of substantial innovation in respect of the mainstream development model.[39] As acknowledged by Pope Francis himself, in fact, the adoption of the 2030 Agenda for sustainable development "is an important sign of hope"[40] and the transformation of such hope into reality could greatly benefit from the cutting-edge message and ecologically sustainable vision brought about by the encyclical letter.

Notes

1. Robert Goodland, "The Concept of Environmental Sustainability," *Annual Review of Ecology and Systematics*, 26 (1995), 1–24, 21.
2. Herman E. Daly, *Steady-State Economics. The Economics of Biophysical Equilibrium and Moral Growth* (San Francisco, CA: W. H. Freeman and Co., 1977), p. 8.
3. On this issue, see, e.g., Nicholas Georgescu-Roegen, *The Entropy Law and the Economic Process* (Cambridge, MA: Harvard University Press, 1971); Nicholas Georgescu-Roegen, "The Entropy Law and the Economic Process in Retrospect," *Eastern Economic Journal*, 12 (1986), 3–25. The issue of the economic relevance of the so-called "entropy law," however, is neither plain nor unquestioned. In this sense, see, e.g., Malte Faber, Horst Niemes, and Gunter Stephan, *Entropy, Environment and Resources* (Berlin: Springer-Verlag, 1987); Herman E. Daly, *Georgescu-Roegen versus Solow: Stiglitz*, presentation at ISEE Conference, Boston, August, 1996; Robert Ayres, "Eco-thermodynamics: Economics and the Second Law," *Ecological Economics*, 26 (1998), 189–209.
4. United Nations Commission on Environment and Development, *Our Common Future* (Brundtland Report) (Oxford: Oxford University Press, 1987).
5. Robert M. Solow, "The Economics of Resources or the Resources of Economics," *The American Economic Review*, 64 (2) (1974), 1–14; Joseph Stiglitz, "Growth with Exhaustible Natural Resources: Efficient and Optimal Growth Paths," *Review of Economic Studies*, 41 (1974), 123–38; Joseph Stiglitz, "A Neoclassical Analysis of the Economics of Natural Resources" in V. Kerry Smith (ed.), *Scarcity and Growth Reconsidered: Resources for the Future* (Baltimore, MD: The Johns Hopkins University Press, 1979).
6. Georgescu-Roegen, "The Entropy Law and the Economic Process in Retrospect," 15.
7. Ernst. F. Schumacher, *Small Is Beautiful. A Study of Economics as if People Mattered* (New York, NY: Vintage Books, 2011), pp. 16–18.
8. Barry Commoner, *The Closing Circle. Nature, Man and Technology* (New York, NY: Knopf, 1971), p. 185.
9. Zygmunt Baumann, *Liquid Times. Living in an Age of Uncertainty* (Cambridge: Polity Press, 2007), p. 11.
10. On this issue, see, e.g., Steve Vanderheiden, "Rethinking Environmentalism: Beyond Doom and Gloom," *Global Environmental Politics*, 11 (1) (2011), 108–13.
11. Daniel P. Castillo, "Integral Ecology as a Liberationist Concept," *Theological Studies*, 77(2) (2016), 353–76, 354.
12. For a detailed analysis of the various components of the integral ecology, see Massimiliano Montini and Francesca Volpe, "In Praise of Sustainability: the Encyclical Letter Laudato Si' and its Legal-Economic Implications," *Italian Yearbook of International Law* (forthcoming).
13. Sean McDonagh, "Laudato Si' Extends Concern to All Creation," *National Catholic Reporter*, July 29–August 11, 2016.
14. More extensively on this issue, see Celia Deane-Drummod, "Laudato Si' and the Natural Sciences: An Assessment of Possibilities and Limits," *Theological Studies*, 77 (2) (2016), 392–415, 405.

15. Christiana Z. Peppard, "Hydrology, Theology, and Laudato Si'," *Theological Studies*, 77 (2) (2016), 416–35, 417.
16. Sean McDonagh, *The Greening of the Church* (Maryknoll, NY: Orbis, 1990), pp. 175–76, cited in Peppard, "Hydrology, Theology," 418.
17. According to Neil Ormerod and Cristina Vanin, the expression "ecological conversion" is due to Pope John Paul II, General Audience Address, January 17, 2001. On this and *amplius* on the issue of ecological conversion, see Neil Ormerod and Cristina Vanin, "Ecological Conversion: What Does It Mean?" *Theological Studies*, 77 (2) (2016), 328–52, 328.
18. Guido Viale, *La Conversione Ecologica. There Is No Alternative* (Rimini: NdA Press, 2011).
19. Ibid., p. 17.
20. Jean-Baptiste de Foucauld, *L'Abondance Frugale: Pour une Nouvelle Solidarité* (Paris: Odile Jacob, 2010), last cover page.
21. Serge Latouche, "Essays on Frugal Abundance. Degrowth: Misinterpretations and Controversies – Part 1 of 4," *Simplicity Institute Report 14c* (2014): http://simplicityinstitute.org/wp-content/uploads/2011/04/FrugalAbundance1SimplicityInstitute.pdf; Serge Latouche, *Petit Traité de la Décroissance Sereine* (Paris: Mille et un Nuits, 2007); Serge Latouche, *Le Pari de la Décroissance* (Paris: Fayard, 2006).
22. Herman E. Daly, *Steady-State Economics. The Economics of Biophysical Equilibrium and Moral Growth* (W. H. Freeman and Co., 1977), p. 2.
23. Herman E. Daly, *From a Failed Growth-Economy to a Steady-State Economy*, USSEE Lecture (June 1, 2009), 3; Herman E. Daly, *Beyond Growth. The Economics of Sustainable Development* (Boston, MA: Beacon Press, 1996), pp. 31ff.
24. André Gorz, Ecologica (Milano: Jaca Book, 2009), p. 37.
25. American Baptist Churches, *American Baptist Policy Statement on Ecology – An Ecological Situational Analysis* (adopted by the General Board of the American Baptist Churches, June 1988): https://acton.org/public-policy/environmental-stewardship/theology-e/american-baptist-churches-policy-statement-ecolog
26. On a similar line of thought, see Klaus Bosselmann, *The Principle of Sustainability. Transforming Law and Governance*, 2nd edn. (London: Routledge, 2016), and in particular ch. 5.
27. Mark Sagoff, The Economy of the Earth: Philosophy, Law, and the Environment (Cambridge: Cambridge University Press, 1988), p. 26.
28. United Nations General Assembly (UNGA), *Transforming Our World: the 2030 Agenda for Sustainable Development* (adopted September 25, 2015). The UN 2030 Agenda, in turn, was issued only a few months before another document due to play a key role at international level: the Paris Agreement on climate change.
29. The seventeen SDGs are as follows: (1) end poverty in all its forms everywhere; (2) end hunger, achieve food security and improved nutrition and promote sustainable agriculture; (3) ensure healthy lives and promote well-being for all at all ages; (4) ensure inclusive and equitable quality education and promote lifelong learning opportunities for all; (5) achieve gender equality and empower all women and girls; (6) ensure availability and sustainable management of water and sanitation for all; (7) ensure access to affordable, reliable, sustainable, and modern energy for all; (8) promote sustained, inclusive, and sustainable economic growth, full and productive employment, and decent work for all; (9) build resilient infrastructure, promote inclusive and sustainable industrialization and foster innovation; (10) reduce inequality within and among countries; (11) make cities and human settlements inclusive, safe, resilient, and sustainable; (12) ensure sustainable consumption and production patterns; (13) take urgent action to combat climate change and its impacts; (14) conserve and sustainably use the oceans, seas and marine resources for sustainable development; (15) protect, restore, and promote sustainable use of terrestrial ecosystems, sustainably manage forests, combat desertification, and halt and reverse land degradation and halt biodiversity loss; (16) promote peaceful and inclusive societies for sustainable development, provide access to justice for all and build effective, accountable, and inclusive institutions at all levels; and (17) strengthen the means of implementation and revitalize the Global Partnership for Sustainable Development.
30. Bosselmann, *Principle of Sustainability*, pp. 76–8.

31. The MDGs, which expired at the end of 2015, comprised eight goals, each of them articulated into a variable number of targets. The goals were as follows: (1) eradicate extreme poverty and hunger; (2) achieve universal primary education; (3) promote gender equality and empower women; (4) reduce child mortality; (5) improve maternal health; (6) combat HIV/AIDS, malaria and other diseases; (7) ensure environmental sustainability; and (8) develop a global partnership for development.
32. www.un.org/sustainabledevelopment/development-agenda/
33. United Nations General Assembly (UNGA), Transforming Our World: the 2030 Agenda for Sustainable Development, Preamble.
34. On the role of regulation for the promotion of sustainability, see Massimiliano Montini and Francesca Volpe, "Rethinking Regulation for Promoting an Ecologically Based Approach to Sustainability," *C-EENRG Working Papers*, 2016–5.
35. UNESCAP Keynote address by Dr. Shamshad Akhtar, *Rule of Law to Support the Post-2015 Development Agenda and Sustainable Development Goals* (May 19, 2015): www.unescap.org/speeches/rule-law-support-post-2015-development-agenda-and-sustainable-development-goals
36. It should be highlighted, however, that the Sustainable Development Goals are not legally binding per se; therefore, they are not directly enforceable.
37. On this theme, see, e.g., Denis Edwards, "'Sublime Communion': The Theology of the Natural World Laudato Si'," *Theological Studies*, 77 (2) (2016), 377–391; Celia Deane-Drummond, "Laudato Si' and the Natural Sciences: An Assessment of Possibilities and Limits," *Theological Studies*, 77(2) (2016), 392–415.
38. Christopher Hrynkow, "The Pope, the Planet, and Politics: A Mapping of How Francis Is Calling for More Than the Paris Agreement," *Journal of Church and State Advance Access* (July 8, 2016), 12.
39. On the shortcomings of the SDGs, see, e.g., Massimiliano Montini and Francesca Volpe, "Sustainable Development Goals: Much Ado about Nothing?" *Environmental Liability*, 4 (2015), 141–47.
40. Speech of Pope Francis to the United Nations (September 25, 2015): http://w2.vatican.va/content/francesco/en/speeches/2015/september/documents/papa-francesco_20150925_onu-visita.html

4

Alter-Ecologies: Envisioning Papal and Ecomodernist Nuclear Energy Policy Futures

Vincent Ialenti

In 2015, two styles of ecological thought were presented as alternatives to mainstream secular-progressive environmentalism. The first was April's *An Ecomodernist Manifesto* (EM): a vision for pragmatic technological progress as a means for achieving poverty reduction, environmental flourishing, and economic prosperity.[1] The second was Pope Francis's May encyclical *Laudato Si': On Care for our Common Home*, which sees climate change as a moral problem exacerbated by consumerism, technocentrism, greed, and individualistic self-absorption. Many have noted differences in these texts' moral compasses and worldviews.[2] Yet much insight can be gleaned from focusing on key ways in which EM and *Laudato Si'* are similar – how they both grapple with questions of how to pursue meaningful lives, how to reckon uncertain futures, and the promises and perils of high technologies such as nuclear energy. To this end, this chapter examines EM and *Laudato Si'* as *alter-ecologies*: alternative calls to re-imagine environmentalism's future ideal worlds. It argues that putting these alter-ecological texts into dialogue with one another – exploring ways in which their visions are complementary versus competitive with one another[3] – can help reveal what is really at stake, unique, and/or not-so-unique about each. What results is a comparative analysis, anthropological in spirit, grounded on sympathetic readings of two subtle visions for technological and environmental futures.[4]

This chapter is organized as follows. First, it reviews ways in which the outlooks of *Laudato Si'* and EM on modernization, markets, technology, technocracy, international law, and human–environment relations are competitive with one another. Second, it describes ways in which the outlooks of *Laudato Si'* and EM on interdisciplinarity, collaboration, "big-picture" thinking, and more holistic forms of environmental progress are complementary in how they pique alter-ecological imagination about the future.[5] Third, it juxtaposes *Laudato Si'* and EM to help think about nuclear power as a means for achieving environmental, innovation, and policy alleviation goals.[6] The chapter concludes by showing how *Laudato Si'* and EM evoke two different visions for climate-friendly energy policy futures. Envisioning possible alternate futures such as these is an essential exercise as twenty-first century values, lifestyles, environmentalisms, and technological circumstances rapidly shift.

PAPAL VERSUS ECOMODERNIST FUTURES

EM and *Laudato Si'* evoke alter-ecological imagination when they criticize paralyzing attitudes of postmodern hopelessness and then advocate that they be abandoned to usher in human and environmental progress. Yet *Laudato Si'* and EM have very different understandings of hopelessness and progress. EM rejects Left melancholy[7] by embracing techno-economic modernist progress. *Laudato Si'* rejects postmodern nihilism and shallow materialism by embracing Christian care for creation and moral progress. They do this in ways that compete with one another.

Laudato Si' and EM both stress how the current environmental crisis cannot be separated from wealth inequalities between the global North and global South, developed and undeveloped countries, the haves and the have-nots. *Laudato Si'*, however, emphasizes the environmental destruction caused by self-centered overconsumption in the developed world. EM laments underconsumption in the developing world. EM notes how the lion's share of future carbon emissions will come not from service economies, but from industrial expansion, infrastructure building, electrification, and other projects increasing qualities of life in the global South. *Laudato Si'* focuses more on ethical principles of ecological debt and common and differentiated responsibilities (LS 52). It advocates, on moral grounds, charitable redistribution of capital and technology to developing nations. EM says that poverty alleviation requires more consumption in the developing world. It concedes that this consumption will – with current levels of technological progress being where they are – inevitably have ecological consequences. This, however, does not mean that international development must slow down. A new coal-fired plant in Bangladesh, EM explains, would release CO_2 and pollute, but would also save lives because economic development extends lifespans (21). EM praises modernization's record in increasing life expectancies, deposing autocratic governments, liberating women from "chattel status," releasing people from hard labor, reducing poverty, and freeing ethnic minorities and kids from oppression (28).

EM does not associate modernization and climate change with irresponsible sin and greed as *Laudato Si'* does. Ecomodernists see climate change as an unintended outcome of "aspirations to a better life" among "people trying to improve their lives and the lives of their children."[8] While *Laudato Si'* does reach out to those who "doggedly uphold the myth of progress" (LS 60), it is more skeptical of modernization's promises. It warns that surrender to technology can dissolve ethical horizons. EM, in contrast, sees moral solutions in technological progress. It insists that innovations in energy technologies, genetically modified organisms (GMOs), carbon capture and storage, agriculture, aquaculture, and global trade can reduce poverty and fight climate change. Many ecomodernists are impatient with

Luddite environmentalists in developed service economies who trash modernization while ungratefully reaping its benefits.[9] *Laudato Si'*, in contrast, points to how so-called progress has flooded the developed world with consumer goods that "baffle the heart" and "prevent us from cherishing each thing and every moment" (LS 222). *Laudato Si'* warns against any technical fix that obscures creation's complex interrelations. No technology can alone open *hearts* in ways that make people genuinely *care* more for creation. Caring hearts can, for *Laudato Si'*, help counter wealthy countries' lack of political will, the irresponsibility of those in power, bad habits such as air-conditioning, shallow materialism, or the moral bankruptcy of profit-oriented cultures of "light-hearted superficiality" (LS 229).[10]

While EM and *Laudato Si'* both deride strictly profit-driven solutions, EM valorizes efficiency and innovation while *Laudato Si'* valorizes charity and humility. Juxtaposing these outlooks with views from the Effective Altruism (EA) movement[11] can help reveal what is at stake with each. Like *Laudato Si'*, EA emphasizes voluntary charitable redistribution from the "haves" to the "have-nots." But, like EM, EA's self-described pragmatic realism focuses less on the purity of the heart's motivations and more on evidence-based reasoning. EA's goal is to define and pursue the most effective net gains to improve the extremely poor's quality of life.[12] As in EM, EA's definition of moral action stresses efficiency and logic more than guilt-inducing self-sacrifice. EM's and the ethics of *Laudato Si'* therefore also diverge in comparable ways to the differences between Mandeville's *Fable of the Bees*[13] and Adam Smith's *Theory of Moral Sentiments*.[14] Like EM, Mandeville does not chastise materialistic desires or self-indulgent pursuits of private gain. He instead emphasizes how private competitive self-interest can reap public benefits. Like *Laudato Si'*, Smith sees individual propriety, pure motivations, charitableness, and everyday good intentions as foundations for virtuous outcomes for society writ large.[15] Smith also, like *Laudato Si'*, stresses the importance of personal virtue even in commercialized liberal societies. Some argue that Smith even thought that commerce could corrupt otherwise decent people.[16] Still, EM and Smith both see ethical concerns as immanent only to everyday practical realities. They do not place them in the transcendent realm that *Laudato Si'* posits.[17]

EM advocates a more incremental progress that lauds market-based technology *substitution* (18): it stresses how the environment is helped when wood energy becomes obsolete and is replaced with natural gas or nuclear energy, when lower-yield agricultural techniques become uneconomical and are replaced with higher-yield techniques, or when urban living – made possible by sanitation innovations – reduces land-intensive rural living. This Promethean view couples moral enhancement with technological advancement: new technologies, for EM, can supposedly help us to live better, more efficiently, or more ethically with Earth's ecosystems. This coupling is, in this way, resonant with transhumanism's vision of human enhancement. EM advocates these substitutions as means for restructuring our material conditions to reduce negative systemic consequences for ecosystems and the poor. But it does not, as *Laudato Si'* does, promote any spiritual or Liberation Theology inflected push to dismantle structures of social sin (i.e., consumerism and technocracy) through renewals of human hearts. EM just sees technology as the best means for restructuring society by disincentivizing harms to the environment and the poor.

EM places less faith in human hearts' potential someday to become more charitable. At times it seems resigned to the economistic idea that people cease to abuse the environment *en masse* only when it becomes very easy for them to do so. On these grounds, EM sees efficiency gains and international trade as more realistic means than personal moral contemplation for reducing poverty. *Laudato Si'*, in contrast, seeks a future "full development of humanity" (LS 62) and rejects today's "deified" market's cult of efficiency (LS 56). It criticizes the globalization of an "undifferentiated and one-dimensional paradigm" of technocracy and capitalism that makes the very "idea of promoting a different cultural paradigm inconceivable" (LS 79 and 108). It notes how such a small number of educated wealthy people have acquired an "impressive dominance over the whole of humanity and the entire world" (LS 104). It warns that market-based, instrumentalist, techno-optimist paradigms can deprive us of purpose, meaning, and identity. *Laudato Si'* would be wary of EM's faith in technology substitution. Rather, the Pope argues, we must attune to how "all of us are linked by unseen bonds and together form a kind of universal family, a sublime communion" (LS 89). EM does not go there.

While EM does see poverty as a problem, it does not share the skepticism of *Laudato Si'* on wealth or efficiency-driven technical knowledge. For the Pope, the current crisis requires nothing less than a revolutionary "renewal of humanity" (LS 118). *Laudato Si'* derides how rampant materialism devalues lesser non-human life forms while also seeing human life as not intrinsically special. The Pope critiques how ecology's values are often reduced to calculative financial logics or absorbed by marketing and corporate social responsibility image-building campaigns. *Laudato Si'* rejects how "finance overwhelms the real economy" (LS 109) as political morality is overtaken by simple cost–benefit analyses. Ecomodernists, however, spend less time on these questions. While *Laudato Si'* rejects the "idea of infinite or unlimited growth" (LS 106), EM rejects strict assertions of resource scarcity, framings of economics as a zero-sum game, and 1970s environmentalists' blind faith in Malthusian growth limits (9). Yet EM rejects unfettered free-market neoliberalism, too. Ecomodernists advocate "strong governments to provide social safety nets, social goods like healthcare and education, sensible regulations, infrastructure and to support R&D."[18] What EM rejects is simply the idea that the world's wealthiest would – by appeals to conscience or heart alone – ever redistribute their capital to the "have-nots" on the scale necessary meaningfully to fight climate change and reduce poverty. Ecomodernists thus tout technology innovations' efficiency gains. They proudly call themselves climate pragmatists or ecopragmatists.[19] *Laudato Si'* would caution EM to make sure that this Prometheanism does not descend into all-too-common "unrestrained delusions of grandeur" (LS 114).

EM and *Laudato Si'* have different visions for future human values and environmental progress. *Laudato Si'* describes a time that some call the Anthropocene – the idea that we are in a new geologic epoch in which humans have become geologic agents[20] – as "beginning to look more and more like an immense pile of filth" (LS 21). EM, in contrast, dreams of a future "good" or even "great" Anthropocene in which human ingenuity allows for both prosperity and natural flourishing (31). *Laudato Si'* regrets that "people no longer seem to believe in a happy future" (LS 113). This does not jibe with EM's self-proclaimed "optimistic view toward human capacities and the future" (31). EM does not share the nostalgia of *Laudato Si'* about human values being "swept away" or needing recovery from a lost past (LS 114). EM does not, as *Laudato Si'* does, contain strong language about piling "debris, desolation, and filth" upon future generations (LS 161). That said, EM does evoke desires to return to something resembling past wilderness conditions when it advocates re-wilding or re-greening (14). But achieving those goals, EM concedes, will not be immediate: game-changing technology substitutions may happen in gradual transitions (24). *Laudato Si'*, too, grants that humanity's "path of renewal" may be long (LS 202). But papal calls for urgent moral revolution assume a different tempo than do ecomodernists' incremental technology substitutions.

EM and *Laudato Si'* seek two very different ideal futures for humanity and Earth's environment. For *Laudato Si'*, humanity's job is to responsibly "till and keep" (LS 66) – to cultivate, oversee, and preserve – Earth's environments. It is not to master, dominate, or exploit them. *Laudato Si'* argues that care, love, and humility can help us to serve responsibly as ecosystems' stewards for generations to come. EM, in contrast, emphasizes *sparing* wild natures by simply not using them: innovating so that we can avoid tilling nature in the first place and, as a consequence, can liberate wilderness so that it can keep itself. For EM, this means *separating* humanity from nature by *decoupling* our actions, desires, and well-being from impacts on it (18). This, too, follows a logic of substitution. Moving to cities, as EM sees it, can spare rural fields. GMOs and fertilizers' efficiency gains can intensify agriculture and reduce its land footprints. Nuclear power plants, using fuels dense in energy (22), spare more land areas than do big fields of wind turbines or solar panels. Desalinization lets more rivers flow free. EM sees such technological replacements as detaching natural areas from human economies in ways that help the planet (18).

This emphasis on liberating wild natures from technologically enabled humanity contrasts with the focus of *Laudato Si'* on the interconnectedness of all creation (LS 70, 92, 240). *Laudato Si'* does not advocate separating from nature as EM does. But it does advocate appreciating the grand "variety of things in their multiple relationships" (LS 86). For Pope Francis, that is what connects integral ecologies with social, moral, spiritual, and aesthetic values. The reverential tone of *Laudato Si'* toward nature sets it apart from EM:

> The bishops of Japan, for their part, made a thought-provoking observation: "To sense each creature singing the hymn of its existence is to live joyfully in God's love and hope" (LS 56) / This contemplation of creation allows us to discover in each thing a teaching which God wishes to hand on to us, since "for the believer, to contemplate creation is to hear a message, to listen to a paradoxical and silent voice" (LS 57) / We can say that "alongside revelation properly so-called, contained in sacred Scripture, there is a divine manifestation in the blaze of the sun and the fall of night" (LS 58) / Paying attention to this manifestation, we learn to see ourselves in relation to all other creatures: "I express myself in expressing the world; in my effort to decipher the sacredness of the world, I explore my own (LS 59).[21]

But EM sees spiritual and aesthetic values in nature, too. While EM does not mention care or the heart, it advocates a "deeper emotional connection" to wild natures that the authors say they love (25, 27). That said, EM does not detail how one might achieve this deeper emotional, aesthetic, or spiritual connection to nature in practice. *Laudato Si'* offers more of a compass. For example, the Pope suggests that we must stop demeaning contemplative rest and pause more often to ponder everything's meaning and purpose. While EM does not share the theological pretenses of *Laudato Si'*, it does say that nature can help people "get outside themselves" by putting us in touch with our deep history (25). EM, like *Laudato Si'*, argues that nature warrants conservation on non-utilitarian grounds (26–27). Pope Francis, however, sees solutions in concepts transcending mathematics (LS 11), whereas EM contains no clear critique of the hegemony of numbers. While *Laudato Si'* seeks a fundamentally new "economic ecology" (210), prominent ecomodernists critique how *Laudato Si'*s asceticism, romanticism, and anti-modernism mirrors conventional environmentalist themes of "sinfulness/redemption and end-times doomsaying."[22] Indeed *Laudato Si'* and EM offer two very different visions for how and why nature should be valued, appreciated, and protected.

Yet EM – even as it glorifies innovation and efficiency – is not, to use Pope Francis's terms, without ethical horizons. EM simply insists that people are most likely to spare natural areas not due to spiritual-aesthetic motivations, but because innovation has stripped those areas of their economic use value. Land-intensive suburbanization, it notes, did not decrease in the United States because people realized that bulldozing forests is immoral, but in response to broader socio-economic and demographic shifts. New England's vast forests did not reappear because Americans' hearts changed, but because agricultural techniques became more land-efficient and intensified, and trade patterns changed. The land sparing that substitutions like these can make possible is, for EM, a moral good.

EM, like *Laudato Si'*, strives to usher in future worlds it sees as more prosperous and just. To see this is to see a deep similarity between two very different texts. The next section explores more complementarities such as these. *Laudato Si'* and EM both, for example, see poverty alleviation as an ethical goal. They agree that energy systems must decarbonize to fight climate change. They agree that ocean acidification and ozone depletion are real problems. They both reject climate denialism and critique complacent responses to ecosystem destruction. So while EM and *Laudato Si'* are competitive in contrasts that leading ecomodernists have rightly called striking, there are parallels between their outlooks for the future. These parallels make them alter-ecologies: useful visions for thinking about alternate planetary, technological, and environmental policy futures.

ALTER-ECOLOGIES

Laudato Si' and EM have more in common than is often appreciated. They both provide alternate starting points for thinking about the future of the planet and humanity writ large. As alter-ecologies, they can both help us – to borrow words from anthropologist Ghassan Hage – "dwell in the world in a completely different way from the way we dwell in it at any given moment."[23] They are in these senses complementary despite their contrasting visions.

Laudato Si' and EM are group efforts that herald collaboration, interdisciplinarity, and "big-picture" thinking about the future. EM presents itself as a multi-authored document developed by economists, an ecosystems modeler, environmental activists, public intellectuals, policy experts, a physicist, an Environmental Studies professor, a historian of geography, a philosopher, a filmmaker, and others. *Laudato Si'* presents itself as a single-authored text that reaches across faiths, cultures, fields, and walks of life to develop a more "integral and integrating" humanism (LS 141) in which "no branch of the sciences and no form of wisdom can be left out" (LS 63).[24] Alluding to works from liberation theologian Leonardo Boff and critical theorist Herbert Marcuse, *Laudato Si'* exhibits its breadth when it cites the ninth-century Sufi mystic Ali Al-Khawas, bishops from outside Europe, Eastern Orthodox Patriarch Bartholomew, and philosopher Paul Riceour. *Laudato Si'* was also influenced by Pontifical Academy of Sciences meetings attended by over sixty intellectuals: Nobel prizewinners, climate scientists, and Hindu, Jewish, Muslim, and Protestant leaders. EM is global in focus, too: it supports its claims with examples from Germany, Japan, California, Bangladesh, New England, the Gulf of Mexico, and elsewhere. In a holistic spirit resonant with that of *Laudato Si'*, EM takes into account economic, technological, emotional, social, ethical, and ecological considerations.[25]

Laudato Si' and EM both defy conventional political imaginations. Pope Francis intentionally avoids "entrenching himself in left- and right-wing ideologies."[26] In *Laudato Si'*, sometimes he "defers to scientists, other times he quotes scripture and, still other times, he criticizes the very foundations of economics."[27] *Laudato Si'*, born of a powerful religious institution long criticized by Leftists, contains both pro-life statements and almost-Marxist critiques of capitalism. Sometimes the Pope's words resonate with critiques of technocracy that have influenced science and technology studies (STS).[28] *Laudato Si'* offers a radical critique of the very notion of private property rights, too. Citing Portuguese Bishops, *Laudato Si'* says that the environment cannot be owned absolutely: it is merely "on loan" to each generation (LS 159). Citing Saint John Paul II, *Laudato Si'* maintains that there is always a "social mortgage" on all private property (LS 93). EM never mentions property specifically. But, like *Laudato Si'*, the ecomodernists – who position themselves against both the "apocalyptic green Left" and the "skeptical libertarian Right" – complicate the "decades-old divide between the pro-growth right and the green left."[29] The way in which EM does this might unnerve secular-progressive environmentalists who are skeptical of nuclear power, GMOs, and fossil energy. Yet EM also rejects starry-eyed neoliberalism, does not advocate organized religion, and supports funding for basic technology research. These visions certainly appeal to, say, science-loving American Democrats.

When *Laudato Si'* and EM unsettle political imaginations, they elicit what Hage terms "alter-politics." That is, they evoke passionate imagination of "alternative economies, alternative modes of inhabiting and relating to the earth, alternative modes of thinking and experiencing otherness."[30] Certainly the new economies – of fresh innovations in EM versus greater charitableness in *Laudato Si'* – are worlds apart. But they both call for alternate modes of inhabiting, relating, thinking, and experiencing that expand what anthropologist Vincent Crapanzano calls "imaginative horizons."[31] Hence, they can help us to re-think future worlds' possible ecologies. *Laudato Si'*, for example, envisions future ecologies of accumulated debris, filth, and climate change to warn against complacency with (post)modernity's status quo of consumerism, profit-obsession, and defiling creation. EM, in contrast, envisions future wild environments left untouched thanks to new innovations in energy, agriculture, trade, and other technologies that foster prosperities necessary for humans to abdicate control over their lands. Both EM and *Laudato Si'* intervene in present human mentalities by appealing to what some STS scholars call future "possible states of the world."[32] They both, to use anthropologist Elizabeth Povinelli's terms, can be means for imagining future ecological worlds "otherwise."[33]

Fostering alter-ecological imagination has become critically important. As Crapanzano has noted, we are increasingly bombarded with "all sorts of end-of-the-world scenarios in literature, film, preaching, and more or less sophisticated discussions of technological failure and economic collapse."[34] This happens as millennial and apocalyptic fatalisms, in the United States and elsewhere, increasingly structure social imaginaries.[35] Risk societies, meanwhile, strive to impose order on disorder, quantify the unquantifiable, or plan for the unplannable as the very "unthinkableness" of future dangers becomes a threat.[36] In these contexts, alter-ecological texts such as EM and *Laudato Si'*, which unhinge imaginations by challenging assumptions grounding how we live our lives, become powerful tools for thinking about future worlds. They can contribute to the twenty-first-century task of exploring "healthy and unhealthy modes of response" to environmental crises without "short-circuit[ing] thinking into counterproductive patterns of denial, despair, and desperation."[37] They can also, to use the words of *Laudato Si'*, help us achieve a "broader vision of reality" itself (LS 138). Juxtaposing *Laudato Si'* and EM, both perspective-broadening texts, can help to broaden perspectives even further.

Laudato Si' and EM, as alter-ecologies, can help us to change how we think, act, live, and love. *Laudato Si'* calls us to envision our future selves as others – to imagine becoming people less sinful and greedy, renewed with open hearts. EM calls on us to envision how seemingly limitless possibilities of human ingenuity can lead to technology innovations that can fundamentally change how we inhabit the world. EM and *Laudato Si'*, seeing limitations in present-day knowledge, therefore both offer open-ended perspectives on humanity's adaptivity, malleability, and capacities for change. *Laudato Si'* asserts humanity's potential for inner, spiritual renewal of its hearts. EM asserts humanity's capacities for outer, technological renewal of its circumstances. Both *Laudato Si'* and EM make these claims in open-access venues available to anyone with Internet access. They are available free online, in HTML or PDF formats, and are not restricted by paywalls. Like *Laudato Si'*, EM invites anyone who wishes to read it to take climate change more seriously and consider solutions that lie outside conventional secular-progressive environmental imaginations.[38] As Brendan Foht put it in 2015, while there are "tensions between their positions, the moral seriousness of Pope Francis and the technological ingenuity of the ecomodernists will both be needed to move us beyond the fruitless debates that characterize so much of environmental politics today."

I have witnessed at first hand the power of *Laudato Si'* to evoke alter-ecological imagination. The openness of *Laudato Si'* to humanity's potentials – plus its holistic, global, interdisciplinary spirit – were what inspired Cornell Law School's virtual think-tank Meridian 180 to host a forum on how it could help us to re-think how governance, environment, spirituality, and markets should interrelate in the twenty-first century. I had the privilege of opening up the discussion, which convened a geographer, a philosopher, a corporate lawyer, anthropologists, a literature scholar, legal scholars, and others. Like *Laudato Si'* – which was published in English, French, German, Arabic, Polish, Spanish, and Italian – Meridian 180's global dialogue was multilingual. Contributors could post in English, Korean, Japanese, or Chinese and, within forty-eight or so hours, Cornell Law School postdocs would translate the posts into the three other languages. In our forum, Korea-based anthropologist Amy Levine drew arguments from EM to critique *Laudato Si'*. Levine asked whether it was unrealistic for Pope Francis to advocate poverty alleviation and limits on economic growth in the same breath given that, in practice, the latter relies on the former. I reported on the ideas that emerged from our reflections on *Laudato Si'* for Religious Left Law and for NPR.org.[39]

Meridian 180 members drew on the power of *Laudato Si'* to help re-think relationships between people and planet. UK-based legal scholar Stephen Humphreys, for example, reflected on how appreciation for art, nature, and the little things in life – in *Laudato Si'* and in Heidegger's thought, too – could help us to escape technocentrism's hollowness. Tetsuya Kaida, a Toyota innovator and management consultant in Japan, discussed how being open to change, to human difference, and to sensory enjoyments such as color, taste, or wind could help to recreate human societies. Korea-based legal scholar Jee Hyung Lee wondered why so many well-off people are overworked as their economies bring desperation to the poor and unhappiness to the rich. United States-based legal scholar Yu Xingzhong worried that modernity cultivates scientific–rational–economic–legal people who do not appreciate the emotional, aesthetic, spiritual, or sentimental sides of life. Scholars based in East Asia explored how popular Daoism, Buddhism, and Confucianism could help us learn to respect nature and to reject waste, luxury, and materialistic desires. Cornell anthropologist Tim McLellan described how his Chinese agroecologist colleagues are open to meaning, wonder, and inspiration even when they study landscapes in highly technical ways. Yet, like Levine, not all participants were fully warm to *Laudato Si'*. Cornell literature scholar Karen Pinkus, for example, emphasized the Catholic Church's questionable history of suppressing human difference – describing how Pope Francis's surface-level prose rests on a "bedrock" foundation of ambiguous institutional power. Oxford geographer Peter Wynn Kirby, while warm to the anthropology-esque respect of *Laudato Si'* for diverse knowledge forms, noted the Catholic Church's "hidebound" tendencies and "imperious" history.

What Meridian 180's forum made clear, however, is that *Laudato Si'*, like EM, can help us to envisage alternate alter-ecologies, lifeways, and ideals. The next section takes this complementarity seriously by drawing upon *Laudato Si'* and EM to imagine different nuclear energy futures. I have selected nuclear energy as my case study in light of my own anthropological research interests in nuclear energy expert cultures. That said, *Laudato Si'* and EM could also help one to think otherwise about GMO, aerospace, fossil energy, IT, permaculture, or other sorts of technology futures, too. After the next section about nuclear energy, this chapter concludes by exploring two different climate-friendly energy policy visions that *Laudato Si'* and EM, respectively call for. The texts' powers to help imagine these futures is, I suggest, what makes them alter-ecologies.

ENVISIONING NUCLEAR ENERGY FUTURES

Lauri Muranen – Director of World Energy Council Finland and a founding member of the Ecomodernist Society of Finland – noted at a Cornell University debate called "Is Nuclear Power the Answer to Climate Change?"[40] that, to produce all the energy a Finn or an American would need in his or her lifetime, one would need just one bucket of nuclear fuel. If advanced nuclear reactor innovations come about, just one soda glass would suffice. With oil, one would need hundreds of thousands of liters. EM similarly extols how nuclear energy's fuel density[41] can leave "more room" for future non-human creatures (6). Energy-dense nuclear energy would, after all, require less mining than, say, coal. EM discusses how nuclear power does not leave the large "land and water footprint" that other energy technologies such as waterway-damming hydropower do (24). Muranen also stressed nuclear power's past successes in decarbonizing energy systems in Sweden and France. EM's authors eagerly anticipate, as Muranen did, the arrival of nuclear fusion power and advanced Generation IV fission reactors – speculating, in an alter-ecological mode, about how humanity "can flourish for centuries and millennia on energy delivered from a closed uranium or thorium fuel cycle, or from hydrogen-deuterium fusion" (10). They envision innovations in reprocessing used-up nuclear fuel to make its long-lived wastes more benign. They envision recycling nuclear waste into fuel inputs for future advanced reactors. Optimizations such as these, in EM's view, could someday make nuclear technologies less polluting and more resource efficient.

Like EM, *Laudato Si'* addresses nuclear energy futures. But *Laudato Si'* focuses more on how nuclear energy, like advances in biotechnology and information technology, has given humanity enormous, unequally distributed, power over populations. Pope Francis is wary of how those with power and knowledge control nuclear power. He stresses how it is historically unprecedented that humans have so much power over themselves. *Laudato Si'* sees this consolidated power as posing future risks, given how twentieth-century Communist, Nazi, and totalitarian regimes used new technologies to kill millions. *Laudato Si'* also notes the mass destruction caused by nuclear weapons, which have long been made alongside nuclear power plants. *Laudato Si'* asks, in "whose hands does all this power lie, or will it eventually end up?" and argues that it is "extremely risky for a small part of humanity to have it" (LS 104). This resonates with concerns about how nuclear energy can entangle with what historian Gabrielle Hecht calls techno-political regimes[42] – empowering collaborating technocrats, financial elites, managers, politicians, corporations, engineers, scientists, and governments to oversee reactor projects. Many worry that this can foster regulatory capture, pose revolving-door risks, and organize decision-making power in ways that *Laudato Si'* would see as being unacceptably technocratic. Hence, *Laudato Si'* would urge EM to be more realistic about future nuclear innovations' powers to solve ecological problems without also creating other new problems. EM would urge *Laudato Si'* to be more realistic about whether climate progress is possible without nuclear power, which requires technocratic control to function.

Indeed *Laudato Si'* and EM promote different ideal futures that, with a little interpretation, can help us re-think nuclear energy futures. *Laudato Si'* might caution against using nuclear energy as a Band-Aid fix to feed the technology, profit, and energy addictions that will consume us unless fundamentally overturned. *Laudato Si'* might assert that nuclear technologies, like any others, cannot alone change our hearts, make us happy, help us love, or teach us to appreciate the interconnectedness of all things. The Pope might worry that overzealousness for technological means such as nuclear power could distract us from seeking more meaningful ends – turning our attentions to hollow, instrumental, economistic logics that obscure the more holistic cost–benefit analyses that humanity sorely needs. Nuclear energy, in the view of *Laudato Si'*, distances us from true progress in a non-quantifiable real economy of human flourishing versus suffering. While many would laud how nuclear power creates science, technology, engineering and mathematics (STEM) jobs, *Laudato Si'* would caution against the technocratic worldviews that it cultivates. Pope Francis's meditations on the moral and political content of technology choices in this way recall STS scholar Langdon Winner's thoughts on the political aspects of adopting technologies such as nuclear reactors.[43]

EM can help us to ponder the ethics of potential nuclear energy futures, too. Nuclear power can, for EM, help pour a steady stream of baseload energy into emerging economies like China, to speed up their development. As anthropologist Dominic Boyer has similarly asked, where would the "biopolitical projects of Foucault's modern prisons, factories and schools" be "without the harnessing and transformation of energy into their lighting and electricity, into their heat, even into their bricks and cement?"[44] Views like these recall William Mulholland's early twentieth-century sense that Los Angeles' great population expansion was contingent first upon huge quantities of water being pumped into the area via aqueducts. The idea is that having plentiful essentials like electricity or water flow freely into a human settlement can, like pouring water and fertilizer into a houseplant pot, prompt its growth. EM brings this logic to bear on poverty alleviation goals in ways that recall Barack Obama's 2013 statements about how the United States' Power Africa initiative could provide "light where currently there is darkness" and "the energy to lift people out of poverty":

> Access to electricity is fundamental to opportunity in this age. It's the light that children study by, the energy that allows an idea to be transformed into a real business. It's the lifeline for families to meet their most basic needs, and it's the connection that's needed to plug Africa into the grid of the global economy.[45]

I reflected on Obama's almost ecomodernist sounding words in a 2013 article on "radioactive green" pro-nuclear environmentalisms:

> [For Obama] providing "access" to electricity is to enable "opportunity." It is to lay down a "lifeline" to fulfill "basic needs" or to transform abstract "ideas" into operable things-in-the-world, into "real" businesses. It is depicted as a "connection" that could serve to "plug" a now-unplugged Africa "into the grid of the global economy," casting the "light" of electrical energy as a solution to the problem of Africa's "darkness."[46]

Visions like these see energy as a trigger for future modernization – a prerequisite for developing future businesses, education, subsistence, and trade. This concept of progress grounds EM's humanism and environmental ethic. It is key to what inspired filmmaker and EM co-author Robert Stone in his documentary about nuclear power *Pandora's Promise*. That film asked whether "the one technology we fear most could save our planet from a climate catastrophe, while providing the energy needed to lift billions of people in the developing world out of poverty."[47]

Here, *Laudato Si'* could interject that nuclear energy's negative externalities can, if not studied thoroughly and managed carefully, reduce qualities of life in unforeseen ways. Radioactive residues and high-level nuclear wastes can, it reminds us, harm human health irreversibly for future generations (LS 21). *Laudato Si's* view of property rights would also seem to rule out nuclear energy. With radioactive half-lives extending tens of thousands, hundreds of thousands, and even millions of years into the future, spent nuclear fuels impose hazards on future populations and ecologies that raise intergenerational equity concerns.[48] On this issue, *Laudato Si'* could add that nature, which cannot be owned absolutely, is only on loan from God to the living. Part of humanity's rental agreement is that we must pass it down to our descendants afterwards. So when people leave behind nuclear waste, they are wrecking God's property, which is already, by contract, scheduled to be loaned to future people. Irreversibly harming it for millennia would mean destroying God's and others' property without permission. Indeed, *Laudato Si'* does not laud radioactive wastes' future inheritances.

EM can help us to ponder nuclear waste's long-term futures, too. Nuclear waste has challenged repository safety assessment experts from Finland to Nevada to devise quantitative models of distant future environmental, hydrogeological, and population conditions.[49] This has led experts to extend ancient logics of legal adjudication to augur deep, geological, planetary timescales.[50] Technological innovation in radioactive waste technologies, EM reminds us, could obviate the need for these projects entirely. Until then, these farsighted forecasts, in tandem with texts such as *Laudato Si'* or EM, can help us think about our planetary futures in an alter-ecological spirit. As anthropologist Richard Irvine has noted, it is important to "open to deep time" in this moment that some call the Anthropocene.[51] In an article about *Laudato Si'*, Richard Miller likewise argues that pondering teleological links between Christology, eschatology, and creation can help "[open] our imaginations to the deep future" in a moment of climate change. Catholic thought about hope and responsibility can therefore help to counter, as Miller puts it, today's "restriction of our imaginations to the short term." Hopefully awaiting the coming of God's kingdom can, Miller argues, help us see far future Earth as our common home even across timespans exceeding those of nuclear waste's risks – providing alternatives to common visions of alien futures of death and absence.[52]

I experienced EM's power to pique alter-ecological imagination – and to help think otherwise about nuclear energy's long-term futures – when I wrote about concepts of generation for the *Cultural Anthropology* academic journal's *Lexicon for an Anthropocene Yet Unseen* project. In that article I noted how many of my critical humanist and social scientist colleagues associated nuclear energy's relationship to the

Anthropocene notion with horror and guilt about capitalism, modernity, privilege, and ecological collapse. These colleagues envisioned worst-case scenario futures plausible in an age of nuclear weapons, population growth, mass extraction, biodiversity loss, environmental destruction, and fears of asteroid impacts and pandemic illness. My ecomodernist colleagues who work as nuclear energy experts, on the other hand, rejected Anthropocene melancholy and instead saw nuclear energy as a climate change solution. While affirming the Anthropocene concept's strength, they were more optimistic about innovation, open futures, technology, incremental progress, human agency, and achieving common political ground across ideological lines. For them, nuclear was an energy/climate technology to be lumped with other energy/climate technologies. EM's futures appeared as alter-ecological counterpoints to humanistic Anthropocene debates' gloomy post-Sixth Extinction, *World Without Us*, or future-Earth-as-present-day-Mars futures.[53] They were not rife with post-apocalyptic despair. I described how ecomodernist futures "invoke future generations to evoke action through enthusiasm and anticipation" for nuclear reactor innovations while critical-humanist Anthropocene commentaries "invoke future degeneration to evoke action through horror or guilt."[54]

Yet EM's warmth to future nuclear technologies is not unconditional. For EM, nuclear power is but an imperfect technology that can be further innovated upon to achieve greater decoupling, resource-efficiency, poverty relief, CO_2 emissions reductions, and land sparing in the future. It is valued as a means to grander ends. While EM affirms that wild nature can help us to get outside ourselves, it does not suggest that technology can do the same. Nuclear power, from this angle, is a work-in-progress that will ideally become obsolete someday when substituted by even better technologies of which we cannot yet conceive. EM presents nuclear energy's viability not as intrinsic, but as relative to other current technologies and, therefore, as subject to change. Hence, nuclear energy innovation futures could have an endpoint. That said, many ecomodernists would, as STS scholar Bruno Latour did in an article for the *Breakthrough Institute*, ask us to "love" our technoscientific "monsters" and to care for them "as we do our children."[55] EM shows a deep respect for the technocratic knowledge necessary to make nuclear technologies possible, noting how they can help "humans use our extraordinary powers to shrink our negative impact on nature."[56] Therefore, like *Laudato Si'*, EM can help us think nuclear energy futures otherwise. I conclude by reflecting on how these alter-ecologies evoke two different climate futures that call for two different energy policy futures.

TWO VISIONS FOR CLIMATE-FRIENDLY ENERGY POLICIES

This chapter has examined ways in which the views of *Laudato Si'* and EM on modernization, markets, technology, technocracy, international law, and human-environment relationships are competitive. It has also examined ways in which the view of *Laudato Si'* and EM on interdisciplinarity, collaboration, "big-picture" thinking, and more holistic forms of environmental progress are complementary. Engaging with these tensions and resonances evoked alter-ecological imagination about possible nuclear energy futures. This can help us to envision climate-friendly energy policy futures, too.

Laudato Si' seeks alternative normative, theological, and ethical foundations for environmental care, energy policy action, and social justice. For energy policies to be legitimate in Pope Francis's eyes, they must be born of healthier politics, franker, more honest debate, more cooperative global solidarities, new frameworks transcending financial logics, fairer decision-making procedures, and more comprehensive cost–benefit analyses. Institutions must not only regulate negative practices, but also promote positive ones. In practice, this could mean establishing policies to weaken economically powerful special interests' powers to control politicians and information. It could mean nuclear power phase-out policies and pushes for more locally controlled renewable energy technologies such as wind or solar. It could mean requiring that environmental impact assessments are not drawn up only *after* policies, plans, or business decisions are proposed, but that they influence strategizing from the start. It could mean more effective energy efficiency initiatives. It could mean establishing disincentives to change harmful First World energy consumption habits. It could mean personally regulating our own greed through sobriety and humility. The climate-conscious futures of *Laudato Si'* would likely move away from nuclear energy.

Pope Francis seeks more generous international cooperations. *Laudato Si'*, like EM, notes that international climate treaties – which are difficult to enforce and often seek hollow emissions targets – have not been adequate. "Cap-and-trade" regulations, EM and *Laudato Si'* agree, have only gone so far. Yet *Laudato Si'* respects and promotes United Nations (UN) norms like differentiated responsibilities (LS 52). It lauds the Vienna Convention's and Montreal Protocols' ozone layer successes and the Basel Convention's toxic waste successes. It maintains that binding international agreements are necessary to help developing nations fund clean energy projects over, say, coal-fired plants.[57] *Laudato Si'* also advocates decentralized, participatory, community-based control over energy generation. This balance is inspired by the Church principle of subsidiarity, which empowers small-scale local decision-making centers that are backed-up by global institutions and international agreements (12). In this spirit, *Laudato Si'* envisions global governance made through international consensuses between states that cooperate while mutually respecting one another's sovereignty. Indeed, Pope Francis is hopeful that more equitable, just, and inclusive decision-making procedures will result in brighter energy policy futures.

Yet the ideals of *Laudato Si'* – calling for extensive rearrangements in how human societies work – provide only limited guidance for practical near-term energy and environmental policies in the world as it stands. It is not difficult to imagine *Laudato Si'* advocating bans on water privatization or legal redefinitions of rights to water access (LS 28, 30). *Laudato Si'* could also advocate that developed countries band together to release developing countries from foreign debt, given how the former historically owe the latter a large ecological debt (LS 51). *Laudato Si'* could advocate policies incentivizing "business creativity" and locally held "productive diversity" like that in small-scale food production projects – resisting food oligopolies' economies of scale that overpower small holders and foster monocultures (LS 129). Programs funding "new forms of cooperation" and "community organization" could be extended to renewable energy cooperatives, too (LS 179, 181, 182). *Laudato Si'* might also argue that carbon credits programs should be expanded for now, but only as provisional responses that supplement, not distract, attention from grander substantive changes that are sorely needed (LS 171). Schools could mandate civics lessons in environmental education to foster "ecological citizenship" (LS 211). Government programs giving financial aid to the poor could be replaced with opportunities for meaningful dignified work (LS 128) – perhaps in energy- or environment-related sectors. *Laudato Si'* could advocate more investment in ecosystems research, too (LS 42).

These policies could, in theory, be implemented very soon. In practice, however, the likelihood that they could be realized on a global scale necessary to mitigate climate change in any meaningful way now seems extremely low. The broad pushes in *Laudato Si'* for more "transparency" (LS 182) and more "enforceable" global environmental agreements (LS 127) simply seem unrealistic today. Further, tensions between the Pope's critiques of technocracy and critiques of unequal global access to high technologies would be at odds in a scenario in which, say, Bangladesh were to pursue a nuclear power plant. The prospect that highly developed countries would send Bangladesh free renewable energy technologies to make good on their differentiated responsibilities or ecological debt seems nearly impossible today. In short, the strength of *Laudato Si'* is not so much in the short-term realm of the actionable or the practical. It is in presenting future ideals for which we can strive that most people, for now, will unfortunately perceive as largely supererogatory. The strength of *Laudato Si'* is in its alter-ecological imagination's ideals.

EM does not unpack the (im)moral foundations of governance procedures. EM is more pessimistic than *Laudato Si'* about international agreements delivering climate-friendly energy policies. As prominent ecomodernist Michael Shellenberger put it, it "has been obvious since the last global climate meeting in Copenhagen in 2009 that none of the big emitters – the U.S., China, India – would allow the U.N. to dictate its energy futures."[58] Ecomodernists place little faith in supererogatory norms and non-binding international commitments. They argue for what they see as more politically feasible solutions. EM could therefore question the wisdom of EU "cap-and-trade policies," pushes to give the UN greater control over energy policies, European bans on fracking, or federal carbon taxes. EM could cast the United Nations Framework Convention on Climate Change (UNFCCC) or Kyoto-style agreements – which rely on powerful states voluntarily sidelining their economic ambitions to help the global commons – as unrealistic and almost always falling short. Why, EM would ask, design policies that make fossil energy costly through regulation when one could instead design policies that make clean energy cheaper through innovation? EM advocates government funding increases for nuclear energy technology research and development (R&D) on these grounds. So while *Laudato Si'* indicts wealthy countries' greed, obsessions with profits, and short-termism, ecomodernists seek economic prosperity far beyond basic needs. EM notes how wealthy countries have been historically more likely to prioritize environment and health than countries that struggle economically. Pro-growth policies could, in this view, double as pro-environment policies.

EM heralds different forms of policymaking cooperation than *Laudato Si'*. Unlike anti-technocratic Pope Francis, top ecomodernists Shellenberger and Nordhaus have argued that revolving doors and knowledge-sharing collaborations between governments and private-sector insiders can spawn innovations.[59] They have pointed to how the game-changing US shale gas fracking revolution emerged from public–private partnerships between businessman George Mitchell, the US Department of Energy (DOE), Sandia National Laboratories, the DOE and gas industry's joint Gas Research Institute, and past findings from President Gerald Ford's 1976 Eastern Gas Shales Project and the Bureau of Mines. Committed national government support like this has, they note, sparked countless innovations in computers, jet engines, microchips, and nuclear power. It has fostered, in their eyes, historical progress from wood to coal to oil to gas. EM could thus support policies increasing state funding for innovations, reducing red tape, and removing barriers to private–public collaboration between experts. That, EM hopes, would usher in more climate-friendly tomorrows that, in contrast with the vision of *Laudato Si'*, would place nuclear power at the heart of energy policy futures.

EM's proposals, more in sync with today's world as it stands, offer more practical near-term policy guidance. Yet EM is unrealistic in its own ways – not just in its starry-eyed zeal for future technologies, but also in its failure to describe just how the new technology rollouts upon which its good Anthropocene hinges could surmount very real hurdles like, say, the precautionary principle or fossil industries' lobbying power. *Laudato Si'* is less at odds with the precautionary principle. EM also fails to address ways in which industrial technologies have been put to uses that are antithetical to aspirations for better lives – as in, for example, the Nazi war machine. What policies, one could ask EM, must be in place to ensure that the tech R&D ecomodernists advocate does not inadvertently result in innovations serving inhumane ends? While these questions remain unanswered, EM's recommendations are still more implementable now than those of *Laudato Si'*. But perhaps this should be attributed less to EM's Promethean optimism than to its resignation to a vision of human nature that cannot be improved or transcended through moral or spiritual cultivation of hearts. That is, EM appeals to human nature as it is today; *Laudato Si'* to the better beings we can become tomorrow. As a result, EM aims to join current conversations about climate-friendly environmental policies and pull them in certain directions. *Laudato Si'*, in contrast, aims to change these conversations' foundations by transcending them or cultivating them.

Yet both EM's and the divergent policy ambitions of *Laudato Si'* cannot be reduced to mere legal, regulatory, or procedural technicalities. They were not made in narrow information silos. They were born of holistic "big-picture" thinking and interdisciplinary collaboration. They were but parts of broader calls to reimagine environmentalism's future worlds. That is what makes them alter-ecologies. Pondering these possible futures becomes essential as global climate crisis, biodiversity loss, nuclear waste, resource extraction, and pollution challenge entire populations to extend their ecological awareness into multi-decade, centurial, or even multi-millennial timescales. Yet envisioning these futures may feel more difficult than ever after high-profile failures of foresight such as the late-1990s dotcom bubble, the 9/11 terror attacks, the BP Oil Spill, Japan's 3/11 tsunami and nuclear disaster, and the 2007–08 global financial collapse and ensuing recession and European sovereign debt crises. As twenty-first-century values, lifestyles, environmentalisms, and technological conditions shift rapidly, alter-ecological visions such as EM and *Laudato Si'* become powerful tools for imagining and inspiring policies for uncertain tomorrows. Bringing them into closer dialogue with one another helps unpack what is at stake in both crucial visions.

Notes

1. John Asafu-Adjaye et al., *An Ecomodernist Manifesto* (Ecomodernism.org, 2015).
2. Brendan P. Foht, "Two Approaches to Climate Action," *The New Atlantis: A Journal of Technology and Society* (Fall 2015), 56; Mark Lynas, Ted Nordhaus, and Michael Shellenberger, "A Pope Against Progress," *MarkLynas.com* (June 20, 2015): www.marklynas.org/2015/06/a-pope-against-progress/; Sally Vance-Trembath, "Modern Pope: Laudato Si' and the Effort to Reform the Feudal Church," *The Breakthrough*, Summer 2016: https://thebreakthrough.org/index.php/journal/past-issues/issue-6/modern-pope
3. By "competitive" I refer to how various perspectives, answers, strategies, or philosophical bases in *Laudato Si'* versus EM are in tension with one another. By "complementary" I do not mean that the views of *Laudato Si'* and EM are exactly the same, but that one can easily affirm compatible perspectives in *Laudato Si'* and EM that are either mutually confirming or at least not mutually inconsistent. Complementary perspectives can speak beyond or past one another while still being posited harmoniously side by side.
4. My approach to analyzing Christian and secular thought comparatively was inspired by anthropological works such as (a) Myer Fortes' 1959 juxtaposition of the Biblical Job parable, the Oedipus myth, and West African ancestor-worship tales, (b) Jane Guyer's 2007 juxtaposition of macroeconomic and evangelical views of the future, and (c) Hirokazu Miyazaki's 2016 juxtaposition of the Book of Job and the conditions of post-Fukushima Japan. See Meyer Fortes, *Oedipus and Job in West African Religion* (New York, NY: Octagon, 1959); Jane I. Guyer, "Prophecy and the Near Future: Thoughts on Macroeconomic, Evangelical, and Punctuated Time," *American Ethnologist*, 34 (3) (2007), 411–21; Hirokazu Miyazaki, "Hope in the Crack of the Social: Reading the Book of Job in Post-Fukushima Japan" in Ingolf U. Dalferth and Marlene Block (eds.), *Hope: Claremont Studies in the Philosophy of Religion* (Tübingen: Mohr Siebeck, 2016).

5. I support this with examples from a fruitful online conversation about *Laudato Si'* that was organized by Cornell Law School's virtual think-tank Meridian 180 in September 2015. See Vincent Ialenti and Meridian 180, "Toward A Global Intellectual Response to Pope Francis' Environmental Thought," *Religious Left Law*, January 18, 2016: www.religiousleftlaw.com/2016/01/toward-a-global-intellectual-response-to-pope-francis-environmental-thought.html
6. I have selected nuclear energy as a topic because it is the focus of my own anthropological work. I support these points with examples from an article that I wrote for *Cultural Anthropology* academic journal's *Lexicon for an Anthropocene Yet Unseen* project that draws upon EM to re-think nuclear energy and ecological futures. See Vincent Ialenti, "Adjudicating Deep Time: Revisiting the United States' High-Level Nuclear Waste Repository Project at Yucca Mountain," *Science and Technology Studies*, 27 (2) (2014); Vincent Ialenti, "Envisioning Landscapes of Our Very Distant Future," *National Public Radio (NPR.org) – Cosmos and Culture: Commentaries on Science and Society* (September 14, 2014); Vincent Ialenti, "Generation," *Lexicon for an Anthropocene Yet Unseen: Cultural Anthropology Fieldsites* (2016): https://culanth.org/fieldsights/846-generation
7. Wendy Brown, "Resisting Left Melancholy," Boundary 2, 26 (3) (1999), 19–27.
8. Lynas et al., "A Pope Against Progress."
9. Foht, "Two Approaches to Climate Action."
10. Certain Breakthrough Institute thinkers – influenced by and influencing ecomodernist thought – have criticized the Pope's anthropological assumption that human motivations are shallow, irresponsible, or superficial. Sally Vance-Trembath, for example, notes how "*Laudato Si'* in places falls prey to a kind of remnant theological anthropology, which over-physicalizes the negative features of the human person and over-spiritualizes the positive ones." (Vance-Trembath, "Modern Pope.")
11. Peter Singer, "The Logic of Effective Altruism," *Boston Review* (July 1, 2015).
12. To use one of Peter Singer's recent examples, one should reconsider following his or her heart's irrational tug to, say, spend thousands of dollars to help a dying Make A Wish Foundation child momentarily happy and instead use that money to save multiple children's lives in an impoverished area where the donation does more good (ibid.).
13. Bernard Mandeville and Phillip Harth, *The Fable of the Bees* (Cambridge: ProQuest, 2012).
14. Adam Smith and Knud Haakonssen, *The Theory of Moral Sentiments* (Cambridge: Cambridge University Press, 2002).
15. Smith's emphasis on common everyday propriety and moral decency resonates specifically with Pope Francis's advocacy of "simple everyday gestures which break with the logic of violence, exploitation, and selfishness" (LS 230).
16. John Dwyer, *Virtuous Discourse: Sensibility and Community in Late Eighteenth Century Scotland* (Edinburgh: John Donald Publishers, 1987).
17. Pope Francis says "[e]nvironmental education should facilitate making the leap toward the transcendent which gives ecological ethics its deepest meaning" (LS 210).
18. Lynas et al., "A Pope Against Progress."
19. The Hartwell Group, *Climate Pragmatism: Innovation, Resilience, and No Regrets* (2011): https://thebreakthrough.org/blog/Climate_Pragmatism_web.pdf
20. Paul J. Crutzen and Euegen F. Stoermer, "The Anthropocene," *Global Change Newsletter*, 41 (2000), 17–18.
21. I thank Frank Pasquale for pointing me to these poetic lines from *Laudato Si'*.
22. Lynas et al., "A Pope Against Progress."
23. Vincent Ialenti and Meridian 180, "Forum Summary: What Role for Global Intellectuals?" *Meridian 180: Transforming the Transpacific Dialogue* (2011).
24. That said, like EM, *Laudato Si'* is – as Michael Perry has put it – multi-authored "across time, in the sense that *Laudato Si'* aspires to be and claims to be continuous with earlier Catholic social teaching; at this time, in the sense that Francis is relying on a broad range of Vatican-related experts" (personal communication, 2016).
25. That said, EM's inclusiveness shies away from inputs from faith leaders or theologians.
26. Christopher Hyrnkow, "The Pope, the Planet, and Politics: A Mapping of How Francis Is Calling for More Than the Paris Agreement," *Journal of Church and State*, 59 (3) (2016), 377, 378.

27. Vincent Ialenti and Annelise Riles, "Excited About The Pope's Visit? Read 'Laudato Si'," *National Public Radio (NPR) – Cosmos and Culture: Commentaries on Science and Society* (September 26, 2015).
28. See, for example, Frank Fischer, *Technocracy and the Politics of Expertise* (Newbury Park, CA: Sage, 1990).
29. See Lynas, Shellenberger and Nordhaus 2015; and Foht, "Two Approaches to Climate Action," respectively.
30. Ghassan Hage, *Alter-politics: Critical Anthropology and the Radical Imagination* (South Carlton: Melbourne University Press, 2015).
31. Vincent Crapanzano, *Imaginative Horizons: An Essay in Literary-Philosophical Anthropology* (Chicago, IL: University of Chicago Press, 2004).
32. Michel Callon, Pierre Lascoumes, and Yannick Barthe, *Acting in an Uncertain World: an Essay On Technical Democracy* (Cambridge, MA: MIT Press, 2009).
33. Elizabeth A. Povinelli, "Geontologies of the Otherwise," *Theorizing the Contemporary, Cultural Anthropology* (January 13, 2014): https://culanth.org/fieldsights/465-geontologies-of-the-otherwise
34. Crapanzano, *Imaginative Horizons*, p. 179.
35. Kathleen Stewart and Susan Harding, "Bad Endings: American Apocalypsis," *Annual Review of Anthropology*, 28 (1999), 285–310.
36. Ulrich Beck, "The Anthropological Shock: Chernobyl and the Contours of the Risk Society," *Berkeley Journal of Sociology*, 32 (1987), 153–65.
37. See Byron Smith, "Doom, Gloom and Empty Tombs: Climate Change and Fear," *Studies in Christian Ethics*, 24 (1) (2011), 77–91, 77.
38. Of course, *Laudato Si'* and EM differ in genre. *Laudato Si'* is explicitly a letter to "every person living on this planet." EM, while implicitly addressed to anyone who wishes to download it, is not a letter. It is a manifesto, a profession of principles, a mission statement, or a creed for ecomodernist thinkers. EM also does not have *Laudato Si'*s binding authority. *Laudato Si'*, as an encyclical from the Holy See, requires Catholics to adhere to it. Among Catholics, as Christopher Hrynkow has noted, only an "informed conscience" (6) which has thoroughly reflected upon the minutiae of *Laudato Si'*, can reject its teachings. This, he explains, raises questions about the moral bases from which self-described Catholic politicians such as Republicans Jeb Bush and Rick Santorum criticized the Pope's authority to write *Laudato Si'* before its release (4–6). See Christopher Hrynkow, "The Pope, The Planet, and Politics: A Mapping of How Francis is Calling for More than the Paris Agreement," Journal of Church and State, 59 (3) (2017), 377–408.
39. See Ialenti and Meridian 180, "Toward a Global Intellectual Response," and Ialenti and Riles, "Excited About the Pope's Visit?"
40. Lauri Muranen and Daniel Kammen, *Lund Critical Debate: Is Nuclear Power the Answer to Climate Change?*, Klarman Auditorium, Cornell University, May 3, 2016.
41. Developing the notion of energy density, ecomodernists note how "long-term processes of decarbonization, moving from fuels that are dirtier, more diffuse, and less reliable to fuels that are cleaner, denser, and more useful economically has been strongly correlated with rising prosperity, better human health, and a cleaner environment all over the world." See Ted Nordhaus, "Fracking's War on Coal: Why Tech Innovation Matters Far More to the Environment than Pollution Regulations," *The Breakthrough*, June 4, 2014.
42. Gabrielle Hecht, *The Radiance of France: Nuclear Power and National Identity after World War II* (Cambridge, MA: MIT Press, 1998).
43. Langdon Winner, *The Whale and the Reactor: A Search for Limits in an Age of High Technology* (Chicago, IL: University of Chicago Press, 1986).
44. See Dominic Boyer, "Energopolitics and the Anthropology of Energy," *Anthropology News*, 52 (5) (2011), 5–7, 5.
45. "Obama Unveils $7bn Africa Power Initiative," *Financial Times*, June 30, 2013.
46. Vincent Ialenti, "Nuclear Energy's Long Now: Intransigent Wastes and Radioactive Greens," Forum on Energy and Social Theory in *Suomen Antropologi: The Journal of the Finnish Anthropological Society*, 38 (3) (2013).
47. Robert Stone (dir.), *Pandora's Promise* (Impact Partners and CNN Films, 2013).

48. Kristin Shrader-Frechette, "Mortgaging the Future: Dumping Ethics with Nuclear Waste," *Science and Engineering Ethics*, 11 (2015), 518–20.
49. See Ialenti, "Envisioning Landscapes," and Vincent Ialenti, "Death and Succession Among Finland's Nuclear Waste Experts," *Physics Today*, 70 (10) (2017), 48: http://physicstoday.scitation.org/doi/10.1063/PT.3.3728
50. See Ialenti, "Adjudicating Deep Time."
51. Richard Irvine, "Deep Time: An Anthropological Problem," *Social Anthropology*, 22 (2) (2014), 157–172.
52. See Richard W. Miller, "Deep Responsibility for the Deep Future," *Theological Studies*, 77 (2) (2016), 436–465.
53. Elizabeth Kolbert, *The Sixth Extinction: An Unnatural History* (New York, NY: Henry Holt and Co., 2014); J. Jones, "No Life on Mars: An Eerie Foretaste of Earth's Future?" *The Guardian*, August 31, 2012; Alan Weisman, *The World Without Us* (New York, NY: Thomas Dunne Books/St. Martin's Press, 2007).
54. See Ialenti, "Generation," *Lexicon for an Anthropocene Yet Unseen: Cultural Anthropology Fieldsites* (2016).
55. Bruno Latour, "Love Your Monsters," *The Breakthrough*, Winter 2012: https://thebreakthrough.org/index.php/journal/past-issues/issue-2/love-your-monsters
56. Breakthrough Institute, "Breakthrough Dialogue: A Good Anthropocene," *The Breakthrough*, July 14, 2015.
57. Hrynkow, "The Pope, The Planet," 20.
58. Michael Shellenberger, "The Climate War is Over," *USA Today*, December 2, 2015.
59. See Michael Shellenberger and Ted Nordhaus, "Uniting a Fractured Republic: Pragmatism, Innovation, and the Shale Gas Revolution," *The Breakthrough*, 2016: https://thebreakthrough.org/index.php/voices/michael-shellenberger-and-ted-nordhaus/uniting-a-fractured-republic.

PART II

The Philosophy and Methodology of *Laudato Si'*

5

The Other Seamless Garment: *Laudato Si'* on the Human Relationship to Created Nature

Mark Shiffman

> Our faith seeks to understand the entire cosmos and everything in it as a seamless garment of God's vast creation.[1]

Beginning in 1971, the phrase "seamless garment" (drawn from the *Gospel of John*, 19:23) was used by Pax Christi founder Eileen Egan to characterize and highlight the unity of Catholic teaching about the respect for life – a unity that cut across and challenged the opposing political loyalties of American Catholics. Joseph Cardinal Bernardin adopted the phrase as an equivalent to his "consistent ethic of life" – a description, which he began to elaborate in 1983, of the deeper moral teachings of the Church underlying opposition to such specific evils as abortion, euthanasia, the death penalty, and nuclear war.[2] By 1998, the United States Council of Catholic Bishops (USCCB), invoking Pope John Paul II's *Evangelium Vitae*, would say:

> Adopting a consistent ethic of life, the Catholic Church promotes a broad spectrum of issues "seeking to protect human life and promote human dignity from the inception of life to its final moment." Opposition to abortion and euthanasia does not excuse indifference to those who suffer from poverty, violence and injustice. Any politics of human life must work to resist the violence of war and the scandal of capital punishment.[3]

In *Laudato Si': On Care for our Common Home* (LS), Pope Francis begins to articulate the outlines of a comparable unity of teaching regarding the human relationship to the natural created world. Like the Church's teaching on life issues, this teaching about our relationship to nature cuts across issues that are not usually treated as interrelated and that appeal to divergent political parties and sensibilities. As he emphasizes from the first pages of the encyclical, while this teaching has developed consistently through the last half-century of papal writings, its themes have also been explored in the reflections of "numerous scientists, philosophers, theologians and social organizations [as well as] other churches and Christian communities – and other religions" (LS 7). As with respect for life, the Catholic Church can offer a teaching on the human relationship to nature that unifies and clarifies concerns shared by all and articulated in part by other communities of discourse. Given this parallel, it can hardly be accidental that Pope Francis concludes this ecumenical stretch of the encyclical with a quote from the Ecumenical Patriarch Bartholomew, who, in calling attention to "the ethical and spiritual roots of environmental problems," speaks of our intimate involvement in "the seamless garment of God's creation" (LS 9).

In order to understand the rhetorical strategy of Pope Francis's ecumenical appeal, and thus to interpret *Laudato Si'* coherently, the reader must begin by recognizing the Catholic understanding of reason that lies behind the document. Its arguments can appeal both to other believers and to non-believers, because it operates on the assumption that we live and think within a created order; this order provides the ultimate standard and guide for our rational apprehension of nature, both human and non-human. This natural order is one that humans do not construct but rather seek always to perceive more adequately. While this belief in created order is ultimately theological, it implies that the shared order provides a basis for rational conversation, for an engagement that can occur *on the level of philosophy* before having to continue on to the theological questions that inescapably follow. Just as the ethic of life presupposes a natural moral law, structured into our existence by the Creator but accessible to rational comprehension by persons of good will, so also what we might call a "consistent ethic of nature" presupposes an integrity of the given natural order of beings – an integrity that we see more readily when we recognize nature as a gift of a good Creator, but that we can also apprehend when our wills and intellects stand in the right relationship to nature itself. Perhaps more pertinent to Pope Francis's strategy, when we stand in the wrong relationship to nature, we can begin to see the inadequacy of our faulty interpretation of its order: misgivings about the consequences of our actions can lead us to question the attitudes that guide them, and thence to question in turn the faulty interpretation that justifies those actions and attitudes. Rational consideration of the flaws in our relationship to nature can lead us to a more adequate apprehension of its true integrity, and perhaps even to an apprehension of its source in the Creator.

Accordingly, at the heart of *Laudato Si'* (in Chapter Three) Pope Francis offers a critical examination of our dominant modern relationship to nature and its flaws, in the hope that by making the terms and tendencies of this relationship explicit we can reflect critically upon it in a coherent way. In language that has become familiar in contemporary educated discourse, he describes the overall contour of this flawed relationship as "the technocratic paradigm" (LS 101 ff.). The paradigm-critique that Pope Francis elaborates proceeds for the first half of this chapter without any reliance on theological argument, before turning to reflection on the corrective resources available from the Christian tradition. Thus, to clarify how the encyclical challenges and offers guidance to all good-willed readers and interlocutors beginning from a strictly philosophical point of purchase, we would do well to begin from the center: the critique of the technocratic paradigm.

DOMINANT PARADIGM

The use of the term "paradigm" to describe basic structural principles of a whole way of seeing, thinking, and acting first gained currency as a result of Thomas Kuhn's use of the word in *The Structure of Scientific Revolutions* in 1962. The primary sense in which Kuhn used the word was to denote the kind of exemplary experiments or theoretical achievements that serve practitioners of scientific research as "exemplars of what to do, the kinds of questions to ask, successful applications, and 'exemplary observations and experiments.'"[4] (Galileo's use of the telescope to observe and interpret what came to be recognized as Jupiter's moons, for example, provided a paradigmatic procedure that eventually differentiated all subsequent astronomy from the previous tradition.) At the same time, however, Kuhn allowed himself to use the term very loosely, so that one critic of his book differentiated twenty-one distinguishable meanings of the word in the text.[5] This looseness has plagued the term ever since, to the point where "paradigm" and "paradigm-shift" have been overused to the point of banality. Our first task, then, should be to clarify how and why Pope Francis uses the word "paradigm" (in Italian, *paradigma*) to characterize "the human roots of the ecological crisis" (LS 101).

To begin with, it would be easy to assume that Pope Francis is calling for a "paradigm-shift" or the adoption of a new or different paradigm in place of the dominant technocratic one. In Kuhn's account of the history of science, and in the more general trends of intellectual history that have taken inspiration from his work, the assumption is that all thought and interpretation operate on the basis of paradigms, and that what we continually see in history is the replacement of one paradigm by another. In fact, Pope Francis only ever uses the word "paradigm" to refer to the deep structure of the relationship to nature he is criticizing.[6] It is possible that the paradigm-structure of our relationship to nature is itself part of the problem.

The text introduces its consideration of the technocratic paradigm by noting that, if we look to the human origins of the ecological crisis, we find "a way of understanding human life and action that has gone astray and that contradicts reality to the point of ruining it."[7] For the sake of reflection on this "way of understanding," Pope Francis proposes concentrating on the technocratic paradigm, which he describes as "dominant." Alongside this paradigm, he proposes that we reflect on the place occupied in the world by the "essere humano," the human being or the being of humanity, and the action that properly belongs to this being. The suggestion, it seems, is that the ensuing reflections will set the errant understanding of our relationship to nature alongside a better and truer account of our more fitting relationship: the paradigm that dominates our relationship will be challenged by contemplation of our right relationship.

The next movement of the text (LS 102–5), before turning to focus on the technocratic paradigm itself, takes up in a preliminary way our relationship to the natural world through the power of technology, by means of which, from our beginnings, humans have transformed nature for useful ends. This activity "expresses the striving of the human soul to overcome certain material conditions," which belongs inherently to our God-given creativity.[8] Nevertheless, the exercise of this power has reached such a scope and extent that it has brought humanity to a point of crisis. Pope Francis offers a characterization of the crisis in terms inspired by Romano Guardini: "Never has humanity had such power over itself, and nothing guarantees that it will be used well, especially if one considers the way it is currently put to use . . . because the immense growth of technology has not been accompanied by development of the human being in the dimension of responsibility, values and conscience" (LS 104–5). This lack of an ethical formation commensurate to our technological power is rendered more dangerous by the way in which this power is concentrated. The knowledge and economic sway of a few provides them with "a striking dominion over the human race and the whole world," and thus even greater temptation to abuse that power as well as greater insulation from the consequences of that abuse.

If this were the root and extent of the problem, then we might look simply to economic and political reform, as well as better formation of conscience and moral understanding, to remedy the crisis. As indispensable as all these may be, Pope Francis has placed this more obvious analysis where it is in the text precisely to emphasize that it is not in itself the most fundamental or profound aspect of our deformed and deforming relationship to the natural world. On this point he is emphatic: "The fundamental problem is a different and even more profound one: the way in which humanity has, in fact, taken up technology and its development *in keeping with a homogeneous and one-dimensional paradigm*" (LS 106, emphasis in original). The problem is not simply that the extent and scope of our power have outstripped our wisdom and virtue. The problem rather is our fundamental understanding of what the power that we are developing is, over what we develop it, and how and why we develop it. Since such power is a human attainment, and its development is a human activity, the root problem turns out to be at the same time a faulty sense of who and what we are as human beings – a faulty sense that both grounds and is reinforced by our technocratic relationship to nature. The self-understanding cultivated in us by the dominance of the paradigm over our minds, and the stimulus to action it implants in our wills, undermine cultivation of the wisdom and virtue that would be necessary to counteract its harmful distortions.

CONTROL VERSUS REALITY

This faulty self-understanding is precisely what the text focuses on in the next few sentences:

> Such a paradigm exhibits a conception of the [human being as a] subject who, through a logico-rational procedure, progressively comprehends/gathers up and in

> this way possesses the object it finds outside itself. This subject manifests itself in consolidating the scientific method and its experimentation, which is already manifestly a technique of possession, domination and transformation. It is as if the subject found itself faced with an unformed reality, entirely available for manipulation. (LS 106)

While human beings have always exercised their creativity by transforming nature to useful ends, the enormous power over nature we now possess results from a concerted and deliberate project of transformation that has the character of possession, domination, and manipulation, for the sake of which we have elaborated and consolidated a body of procedures and techniques that we call scientific method. If modern methodological science is the task in which humans fulfill their potential simultaneously as knowers of nature and, by virtue of the same activity, its masters and possessors, then it is this relationship to nature that implicitly communicates to us and forms us in our sense of who we are as human beings: subjects who act to attain power over nature.

This claim – namely, that the relationship to nature structured in the practice and institutionalization of modern scientific method is not only problematic but even "contradicts reality" – will seem extreme and outlandish to most modern readers. This should not be at all surprising, on the premise that their minds are caught within the confines of the paradigm. We must note, however, first, that the claim is far from unprecedented and, second, that its precedents may be found on either end of the political spectrum. The deep critique of techno-scientific knowledge-structures and practices will be familiar to anyone who has read the philosophy of Heidegger or those influenced by it, as well as to anyone familiar with eco-feminist thought or other Green movement writings on the left. It will also, however, be familiar to agrarian traditionalists on the right (famously described by Rod Dreher as "Crunchy-cons"), as well as Thomistic or Aristotelian critics of the modern conceptualization of nature. Indeed, many figures who develop this kind of critique, such as Wendell Berry or Pope Francis himself, are hard to place on a political spectrum at all.

We must emphasize that there is no question here of any religious hostility to science, such as "New Atheists" (usually without scrupulous consideration of the evidence) like to imagine as the norm among believers. Pope Francis is participating in an already existing tradition of philosophical critique, elaborated in philosophical terms, before opening a vista onto the relevance of any theological perspectives. The fact that a similar critique can be found issuing from so many diverse and conflicting philosophical starting-points in itself suggests that there is a real problem in need of adequate articulation. The important thing, of course, is whether the critique carries plausibility.

In fact, establishing the prima facie plausibility of Pope Francis's characterization of scientific method requires little philosophical sophistication. What, after all, is this practice that we call scientific method? The primary foundation of all modern scientific credibility lies in experimentation. For the older view, grounded in Aristotle's thought, nature was characterized by the way in which things, when left to themselves, act "always *or for the most part*."[9] Modern science distinguishes itself by using exactitude to show what must necessarily always follow from what. In the most rigorous cases, experimentation means the establishment of a starting state of exactly defined and measured conditions, into which the experimenter makes an exactly defined and measured intervention, which consistently yields the same exactly measurable final outcome state.[10] In other words, we show definitively that if we can control the initial conditions and the modification of them, we also thereby invariably control the result. While it may be possible to distinguish them analytically, the employment of the criterion of exactitude in such experimental procedures is in practice always and necessarily identical to the employment of the criterion of control, and thus the knowledge obtained is always equivalent to power.

This power, especially when it is the power to unlock, reconfigure, and store up energies hidden within nature's given forms, becomes a power of transformation that can have only very little if any regard for those given forms in which nature orders itself. "It is as if the subject found itself faced with an unformed reality, entirely available for manipulation." As far as the physicist is concerned, the "reality" of things is, in the eyes of experimental method, the energies that can be known only by releasing them from their naturally ordered forms and controlling them. The human being relating in this way to nature is the subject who obtains and exercises this control over "objects"; and these objects are constituted and defined (one might say, extracted) by the subject's action on the malleable world through the interface of experimental method.[11] For modern science, the relationship of subject to object is inescapably the relationship of control: to be a knowing subject is to exercise dominion over nature by extracting controllable objects from it.

Of course, not everyone conducts scientific experiments in which knowledge-as-power is attained or reconfirmed (though, at some point in their education, most educated persons are initiated into the conduct of confirmatory experimentation). Within the institution of science itself, much research involves fitting collections of measurable data into models of how one part of nature works. Here, too, extraction of objects of study according to the demands of quantification focuses attention on very limited aspects of the whole reality, and it is easy to confuse the model with the reality itself. While many researchers are motivated by curiosity and wonder, and retain a capacity to recognize beauty and mystery in the world that they study, the very form of the paradigm tends to enforce its own demands and shape its own institutional ethos.

The "non-scientists," however, who receive the pronouncements of "science" as an authoritative cultural voice, are even more deeply dominated by the technocratic paradigm and its equation of knowledge and power. The consolidated findings of scientific experiment are considered true by most persons simply because "they work" – that is, the technological products that use these findings for their construction and operation do what they are designed to do (at least on a good day). The paradigm is "technocratic" because it gains its power over the human soul through the manifest technological power that it gives its adherents over the forces of nature.

WITH AND AGAINST NATURE

To illustrate the change that this technocratic paradigm has brought about in our relationship to nature, Pope Francis offers a curious image:

> One finds at all times the intervention of the human being into nature, but for a long time it had the character of companionship, of working further in the direction of the possibilities offered by things themselves – a matter of receiving what natural reality itself allowed, as if it were extending a hand. Now, on the contrary, what is looked for is the extraction of as much as possible from things by forcing human hands upon them, which tends toward blindness to or forgetfulness of the very reality of what is in front of us. In this way the human being and the things have ceased to offer each other a friendly hand, instead becoming adversaries. (LS 106)

This contrast between friendly hand-clasping and violent grasping may seem fanciful (and of course it ignores the sense of nature as threatening that we also find in most cultures), but it expresses a real historical change. If we think of the water-mill – an invention that magnified the productivity of human work considerably, but within limits – we can see not only that it makes use of what nature is already doing without significant impact on the river, but also that it establishes a harmonious relationship between humans and the river basin. Certainly, maintaining that harmony requires the exercise of responsibility; but it is a responsibility that it is within the power of human prudence to exercise. The immense multiplication of power brought about by coal or electricity, on the other hand, demands exactly the kind of violent extraction that Pope Francis implies, and these demands would punish any business that tried to maintain them within limits narrower than those of the competition. Likewise, traditional farming that includes both animals and crops works with the rhythms and cycles of nature by using animal waste for fertilizer and previously harvested fields for pasture, whereas agribusiness that specializes in one product on a massive scale breaks the cycle, so as to "take a solution and divide it neatly into two problems."[12] Pope Francis identifies the refusal to take such circular systems as exemplary for our relationship to nature as one of the characteristic features of what he calls a "throwaway culture" (LS 22).

Pope Benedict XVI makes a similar point when he defends the Christian tradition and the Book of *Genesis* from the charge that, by acting on God's command to "subdue the earth," Christianity introduced into human history the ethic of domination and exploitation. According to its critics, Christianity "transformed all the powers of the universe, which were once our brothers and sisters, into utilitarian objects for human beings, and in so doing … has led them to misuse plants and animals and all the world's powers." Pope Benedict responds:

> The Creator's directive to mankind means that it is supposed to look after the world as God's creation, and to do so in accordance with the rhythm and the logic of creation. The sense of the directive is described in the next chapter of *Genesis* with the words "to till it and keep it" (*Genesis* 2:15). An allusion is made here to the terminology of creation itself, and it signifies that the world is to be used for what it is capable of and for what it is called to, but not for what goes against it.[13]

In this way, Pope Benedict points to the same kind of change from a cooperative relationship to the adversarial one that Pope Francis sees the technocratic paradigm bringing about. He also sees its connection to a loss of respect and regard for the forms of things given by created nature itself: "Previously human beings could only transform particular things in nature; nature as such was not the object but the presupposition of their activity. Now, however, it itself has been delivered over to them *in toto*."[14]

Both Popes, in other words, point to an ethic of conservation and sustainability that is primarily grounded not in moral and legal stipulations, but rather in the sense of a relationship of humility, loving regard, and mutual dependence. While Pope Benedict, engaged in the work of interpreting Scripture, stresses from the outset that this sensibility flows necessarily from the Biblical understanding of creation as gift and responsibility, at this point in his text Pope Francis merely characterizes the phenomenon of the relationship itself, highlighting the change that domination via the technocratic paradigm brings about. Of itself, this paradigm places us into an adversarial relationship to nature characterized by an impulsion to endless extraction. While that extractive relationship and its irresistible excesses appear most visibly in our economic practice (to which we will turn shortly), as we have seen, it is important to recognize that this extractive mode already characterizes the way in which modern scientific experimentation constitutes its objects. The technocratic paradigm makes extraction seem like our normal relationship to nature, because according to the logic of its practice nature really *is* the objects we have extracted from it and the necessary and controllable relationships we have identified among them.

At the heart, then, of our newly adversarial relationship to nature lies this grand project of extracting controllable objects and mapping out their controllable relationships, and then identifying this map with nature itself. When we thus approach nature as "an unformed reality, entirely available for manipulation," the map we construct and with which we overlay it inescapably and even aggressively "tends toward blindness to or forgetfulness of the very reality of what is in front of us," namely the given forms of natural beings. No doubt, this conceptual map gives us new insight into ways in which "everything is interconnected"; but, as a construct, it leads us to lose sight of the integrity with which each natural being presents itself to our attention, the way in which every living being is "good and admirable in itself" and lives out its own proper part in a "harmonious ensemble of organisms" (LS 140).

From this failure of respect for natural form irresistibly follows a failure to respect natural limits. If our knowledge is constituted by discovery of the extent of what we can extract and control, the quest for that knowledge commits us to continual transgression of all limits, as Michael Hanby remarks:

> [S]ince we can discover the ultimate limits of technological possibility only by transgressing the present limits of possibility, the technological paradigm commits us to a perpetual war against the given limitations of nature. This is why the so-called technological imperative, that what can be done must, has proven so difficult to resist. It is built into our notions of reason and nature. From the vantage point of the technological paradigm, to fail to heed this imperative is to forsake reason itself.[15]

It is because this transgressive imperative is built into the technocratic paradigm that it shapes a "way of understanding human life and action" that "contradicts reality to the point of ruining it."

We can see why Pope Francis identifies the technocratic paradigm as the "fundamental problem" in our relationship to nature: the institutionalized practices of modern scientific research and its technological applications shape our dominant understanding of what knowledge and nature are, and thereby render us hostile to any limits on our own mastery that a harmonious relationship with nature would require us to respect. This fundamental dimension of our distorted relationship to nature is intellectually the most difficult to challenge, especially given the way in which technology, which increasingly mediates and obscures our relationship to nature, "tends to absorb everything into its ironclad logic" (LS 108). At the same time, it is not difficult now to see how this dimension of our attitude toward nature grounds the other two major dimensions Pope Francis also challenges us to reflect upon: our economic practice of appropriating and using natural resources, and our struggle with the natural procreative sexual powers of our own bodies. The order and disorder of these three dimensions of our attitude toward nature are intimately bound together as one seamless garment.

LIMITLESS ACCUMULATION

Pope Francis himself, still in section 106, passes immediately from characterization of our adversarial relationship to nature to the problem of observing limits in economic life:

> From there one progresses easily to the idea of infinite or unlimited growth, so warmly embraced by economists and theorists of finance and technology. This idea

> rests upon a falsehood concerning the infinite availability of the planet's goods, which leads us to press it to the limit and beyond the limit.

Comment upon this passage could itself fall prey to infinite growth; let us limit ourselves to its connections to the technocratic paradigm as we have already elaborated them.

It is, in the first place, not difficult to verify the dominant tendency in contemporary public culture to identify the health of a nation's economy with its rate of growth; any mainstream media discussion of any nation's economy will bear this out. The most common indicator of growth is production. Economists will also assess the sustainability of patterns of growth in the near term, which generally means at most about a decade into the future, and more commonly only a few years. On the face of it, then, economists do not explicitly assume the possibility of infinite growth, since they tend to limit predictions to the near term. The logic of the growth standard, however, is obvious: the goal of economic growth is, in principle, a perpetual goal. Reactions to the economic vision of *Laudato Si'* make it evident that the proposal for purposefully slow or limited growth is anathema to the economist's mind, to say nothing of the notion of a stable state.[16] If there is *never* a time at which minimal or zero growth is an acceptable economic prescription, this amounts to assuming infinite growth as our economic goal.

Production requires resources. The projection of unending growth implicitly commits us to infinite resource consumption, continually pressing beyond sustainable limits. This projection of infinite growth thus lies at the heart of a constructed model that "contradicts reality to the point of ruining it." Infinite growth is mathematically possible, and therefore imaginable, in a model that assumes resources. In reality it is not possible because, even granting a certain proportion of renewable resources (and the inclusion of services as well as products in measures of growth), non-renewable resources will be exhausted and, as they become scarce, the welfare of particular human beings will be sacrificed in the frenzy to extract more of them.

The entanglement of economics in the technocratic paradigm, however, often makes it difficult to see such problems clearly. The prestige of the economic profession as such derives largely from its resemblance to the natural sciences, which it achieves by limiting its focus to quantifiable objects, and above all money.[17] Technocratic economics thus inevitably "tends toward blindness to or forgetfulness of the very reality of what is in front of us," inasmuch as the tangible degradations of the natural and human ecologies that may result from profitable ventures often cannot enter into the accounting system in a way that commands the authority of economic science. The national scope of political authority reinforces this technocratic tendency: high-level decisions to be imposed on local communities gain much of their authority from applying the calculations of experts "regulating from nowhere."[18]

This technocratic ethos leads to what Wendell Berry describes as "an economy based upon several kinds of ruin" which "substitutes for the real economy, by which we build and maintain (or do not maintain) our household, a symbolic economy of money, which in the long run ... cannot symbolize or account for anything but itself."[19] The science of economics tells us how to sustain growth of the symbolic economy indefinitely, but leaves it to others to figure out how to find the real resources. It treats the world as the technocratic paradigm does, as an abstract system of quantifiable relations of cause and effect, from which objects (in this case embodiments of "value") can be constituted by processes of extraction.

Pope Francis recognizes that a healthy "human ecology" requires support for the dignity of the human "vocation to work" (LS 128), which in turn requires attentiveness to and protection of local economies.[20] Economies of scale, driven by excessive enchantment with technological progress and devotion to profit-maximizing or efficiency-maximizing economic models, lead to economic freedom for the powerful and wealthy, but diminishment of freedom for those who lose jobs in the process of consolidation and greater efficiency.

> The mere proclamation of economic freedom, while actual conditions prevent many people from really being able to take part in it and the opportunities for work dwindle, becomes a self-contradictory discourse that brings politics into disrepute. Business activity, which is a noble vocation oriented to producing wealth and bettering the world for all, can be a fruitful way of improving the area in which its activities are located, especially if it is understood that the creation of positions for labor is an indispensable part of its service to the common good. (LS 129)

Economic thought operating in the form of the technocratic paradigm, on the other hand, will look upon workers and localities as resources quantifiable within a model of efficient conversion of raw materials or labor time into value objects. It will look upon the sale of value objects as maximizing profits, usually distributed to distant investors, or in terms of greater efficiency. Such technocratic economics recognizes no compelling stake in the cultivation of human beings and localities when greater efficiency and profit requires treating them as variables in an equation in which substitution is a simple mathematical operation. Again, while many economists undoubtedly approach their work with humane motives and intentions, the sway of the technocratic paradigm has a tendency to undermine the more complete understanding of human dignity and happiness that would be needed to render those intentions effective.

According to Pope Francis, the technocratic misunderstanding of the value of labor results from the loss of "the capacity for contemplation and respect" (LS 127), for seeing the reality of what is in front of us and recognizing its goodness and dignity. As we shall see, it is this contemplative and loving capacity that he will hold out more generally as the alternative to seeing from within the technocratic paradigm. The important point here is that economic thought operating out of the technocratic paradigm will tend, even despite good intentions on the part of its practitioners, to reduce both the earth and human beings to exploitable parts of "an unformed reality, entirely available for manipulation." Pope Francis calls for our economic plans and projects, and so also our economic theory on the basis of which we formulate plans and policies, never to lose sight of the human being's contemplative and respectful regard for natural beings and human persons, on the basis of which we become able to prioritize their healthy relationship in real localities. This is central to what he describes as "integral ecology."

FAMILY ECOLOGY

In the fourth chapter of *Laudato Si'*, Pope Francis turns explicitly to the topic of integral ecology. In the subsection on "Ecology of Daily Life" (LS 147–155), he draws the connection to the third aspect of what we are calling the seamless garment of nature, namely our relationship to our own bodies. He himself makes the connection explicit: "A logic of dominion over one's own body transforms itself into a logic, often subtle, of a dominion over the creation" (LS 155). Explicitly quoting Pope Benedict XVI's reminder that "the human being has a nature that must be respected and cannot be manipulated at one's pleasure" and implicitly invoking Pope John Paul II's principle of "respecting the meaning of one's own body" as "essential to human ecology," Pope Francis goes on to emphasize the need to "appreciate also one's own body in its masculinity or femininity" if one is genuinely to "encounter the other" (LS 155).

Given the focus in this encyclical on "care for our common home," it makes sense that Pope Francis would emphasize this direction of causality: our cultural and political formation promotes a claim of dominion over our own bodies, which shapes our attitude of dominion over nature as a whole. Conversely, in *Amoris Laetitia* (AL), in which the focus is on the theology of family life, he notes that the causal logic can go in the other direction:

> We treat affective relationships the way we treat material objects and the environment: everything is disposable; everyone uses and throws away, takes and breaks, exploits and squeezes to the last drop ... Narcissism makes people incapable of looking beyond themselves, beyond their own desires and needs. (AL 39)

Amoris Laetitia also draws attention to the consumerism necessarily encouraged within the economy of growth, citing it as a corrosive force on the family. In the first place, it "deter[s] people from having children, simply so they can maintain a certain freedom and life-style" (AL 42); and even if they do choose to have children, the "tensions created by an overly individualistic culture, caught up with possessions and pleasures, leads to intolerance and hostility in families" (AL 33).

These two texts, taken together, suggest a circular and mutually reinforcing relationship between our economic conceptions and attitudes and those concerning the body and sexuality, as well as a deep connection of both with the tendencies of our technocratic paradigm for knowledge of nature.[21] The strongest statement of the connection comes in AL 56:

> Yet another challenge is posed by the various forms of an ideology of gender that "denies the difference and reciprocity in nature of a man and a woman and envisages a society without sexual differences, thereby eliminating the anthropological basis of the family. This ideology leads to educational programmes and legislative enactments that promote a personal identity and emotional intimacy radically separated from the biological difference between male and female. Consequently, human identity becomes the choice of the individual, one which can also change over time" ... It needs to be emphasized that "biological sex ('sex') and the socio-cultural role of sex ('gender') can be distinguished but not separated" ... It is one thing to be understanding of human weakness and the complexities of life, and another to accept ideologies that attempt to sunder what are inseparable aspects of reality.

Like the critique of the technocratic paradigm in *Laudato Si'*, this passage turns our attention to a given natural order, and to the human attempt to subject it to the control of our wills by imposing a conceptual structure that falsifies that reality.

The contemporary theoretical and practical assertion of the fluidity and changeability of gender, which Pope Francis singles out here as a form of violence against the natural order, provides a perfect illustration of the seamless-garment problems he is trying to clarify in chapters three and four of *Laudato Si'*. In the teaching of gender fluidity in schools, Pope Francis sees a form of "ideological indoctrination" imposed by wealthier and more powerful segments of humanity on those in need of their financial support for educational institutions.[22] Not only are the economic and sexual transgression of limits in the name of individual freedom of choice connected with one another, but they are underwritten by the technological power over nature possessed by those wealthy and "liberated" portions of humanity. Indeed, under the sway of the technocratic paradigm, our power to reshape the sexual characteristics of the human body dictates what we think we know about the fluidity of gender and sexuality, what we are allowed to say about the parameters of their reality. Gender is something about which we can make up our own minds because reality is something we construct, which we prove by being able to control it.[23] Moreover, the relative lack of evaluative discussion with which transgender medical therapies have been accepted as health measures, and the relative silence in the medical community about the studies suggesting that they are more harmful than beneficial even according to therapeutic criteria, provide a perfect case study of the way in which our enthusiasm for new technological power transgressing previous limits constantly runs ahead of cultivating sound ethical judgment about its exercise.[24]

Failure to recognize the way in which these distortions of our relationship to nature are deeply interconnected partly accounts for the widespread confusion regarding Pope Francis's teachings on sexuality and marriage. How can the Pope who preaches compassion for the marginalized, refuses to greet homosexuals with moral scolding (even bestowing his apostolic blessing upon a gay woman and her spouse), and says "Who am I to judge?" also say that the struggle for same-sex marriage is "not a simple political struggle [but a] destructive attempt toward God's plan" in which "the identity and survival of the family" is at stake?[25] Writing in *Salon*, Nico Lang comments: "The pope's message on trans people boils down the old canard that the Catholic Church has been using against gays and lesbians for decades: 'Love the sinner, hate the sin.'"[26] Pope Francis himself, in the passage from AL 56 quoted above, provides the more accurate explanation: "It is one thing to be understanding of human weakness and the complexities of life, and another to accept ideologies that attempt to sunder what are inseparable aspects of reality." The delusion that we freely construct our own identities in the midst of "an unformed reality, entirely available for manipulation" is an affliction to which the more powerful and wealthy nations are prone by virtue of the economic and technological dominion over nature they have attained; and they are also prone to treat all of humanity as malleable material upon which they are free to impose this ideological form in the name of progress. Unlike Pope Francis, the technocratic-individualist view of freedom tends to equate charity with identity affirmation.

Pope Francis speaks of the "inseparable aspects of reality" sundered by gender ideology and same-sex marriage advocacy in quite traditional natural law terms, which again rest upon acknowledging an order to reality recognizable by reason but not by technocratic objectification. In a 2010 letter to the Carmelite Nuns of Buenos Aires, he encourages them to pray that Argentine legislators will act "according to the natural law and the law that God points out to them," rather than succumb to "the outright rejection of the law of God, engraved also in our hearts."[27] In the same vein, in *Amoris Laetitia* he affirms that "according to the order of creation, conjugal love between a man and a woman, and the transmission of life are ordered to each other" (AL 81). While some progressive admirers of Pope Francis may be inclined to regard such claims as an old churchman's outdated defense of traditional gender roles, this would represent a failure to take seriously the deeper connections he sees in our attitudes toward nature, including the tendency to treat our own bodies as artifacts and to believe that we are free to decide what to make of them and what uses to put them to.[28] In other words, one may justifiably suspect that such a dismissal originates from inside the technocratic paradigm he rejects, which sees nature as raw material and reality as a human construct.

THE SAINT FRANCIS ALTERNATIVE

To be sure, these questions are far more complicated than we can do justice to here. Fully unfolding the breadth and depth of the teachings and giving them clear articulation is an ongoing effort, as is understanding the practical and pastoral implications. But only by recognizing the fundamental connections at the heart of these distorting relationships can we begin to do them justice, and to recognize the coherence of Pope Francis's critique. Seeing that coherence also helps us to recognize the fundamental significance of the alternative Pope Francis proposes as an indicator of the way toward clearer vision.

It is explicitly as the embodiment of a different relationship to nature that Pope Francis invokes his namesake, Saint Francis of Assisi, whose hymn to God's creation gives the encyclical its title. This hymn is the poetic expression of what every aspect of the life of Saint Francis exemplified: an "openness to awe and wonder," a readiness to speak "the language of fraternity and of beauty in our relationship with the world," and a "refusal to turn reality into a mere object of use or of mastery" (LS 11). If we close ourselves off from such receptivity to the inherent goodness and beauty in created nature, "our attitude will be that of masters, of consumers, or of mere exploiters of natural resources, incapable of placing a limit on their immediate interests." This receptivity is what Pope Francis means by contemplation, which, as Cardinal Peter Turkson has emphasized, is a keynote of the encyclical.[29] When, in the final chapter, Pope Francis proposes the model of the saints for cultivation of an "ecological spirituality," he invokes the great Franciscan theologian Saint Bonaventure: "Contemplation is heightened to the degree that the human being feels the effect of divine grace within, or recognizes God in other creatures" (LS 233).

Contemplation, love, and respect for the goodness, beauty, and dignity of created beings, then, is not a different paradigm, but rather an alternative to structuring our relationship to nature by imposing a paradigm. By this receptivity, we begin to overcome the "blindness to or forgetfulness of the very reality of what is in front of us"; we thus allow ourselves to be drawn by the hand nature stretches out to ours into "a fuller vision of reality" into which our partial knowledge of "objects," separated out and constructed within the constraints of methodologies, can be reinterpreted into a whole in which we have a place no longer as a dominating "subject" but as an intimate participant (LS 139). Contemplation allows philosophy to play an integrating role that transcends the limited purview of science, enabling us to recognize, in a way that the objectifying operations of the technocratic paradigm do not, that "reality is greater than ideas" (LS 110). As the quote from Saint Bonaventure suggests, that openness to the greater reality is also what enables us to recognize and accept grace, to experience the natural world as a created order, a divine gift in which the Creator is present to be greeted with love. What we need is not a new paradigm, but rather "a personal conversion" (LS 216).

Notes

1. Ecumenical Patriarch Bartholomew, *In the World, yet Not of the World: Social and Global Initiatives of the Ecumenical Patriarch Bartholomew* (New York, NY: Fordham University Press, 2010), p. 62.
2. See the second paragraph of Bernardin's William Wade Lecture at St. Louis University, "A Consistent Ethic of Life: Continuing the Dialogue," March 11, 1984: www.priestsforlife.org/magisterium/bernardinwade.html
3. "Living the Gospel of Life: A Challenge to American Catholics," *U.S. Conference of Catholic Bishops*, November 22, 1998: www.usccb.org/issues-and-action/human-life-and-dignity/abortion/living-the-gospel-of-life.cfm
4. Ian Hacking, "Introductory Essay" in Thomas Kuhn, *The Structure of Scientific Revolutions* (Chicago, IL: University of Chicago Press, 2012), pp. xxi–xxii.
5. Margaret Masterman, "The Nature of a Paradigm" in Imre Lakatos and Alan Musgrave (eds.), *Criticism and the Growth of Knowledge*, (Cambridge: Cambridge University Press, 1970), pp. 59–90.
6. The one seeming exception is in LS 108, where Pope Francis rejects the possibility of promoting "a different cultural paradigm" in which technology would be reduced to a mere instrument. While he is not explicit on the point, the counterfactual use here is consistent with a reading according to which constructing a paradigm is part of the problem.
7. LS 101 (my translation). The official English translation is quite inadequate here, as in many passages. It speaks of a "way of understanding human life and action that has gone awry, to the serious detriment of the world around us," as though the problem were how the paradigm is deployed rather than a fundamental erroneousness at its foundations. My translation – here, as throughout the chapter – follows the Italian more literally, as do the official French, Spanish, and German translations.
8. LS 102, quoting Pope Benedict XVI, *Caritas in Veritate* (2009), para. 69 (my translation of Italian rendering).
9. Aristotle, *Physics* II.8, 198b35-36.
10. The resort to statistical probabilities in post-Heisenberg quantum mechanics does not fundamentally change this description, except by modifying the parameters of control. Whether "chaos theory" challenges the paradigm of control is a more difficult question.
11. On objectification (specifically as it affects biology rather than physics), see Leon Kass, *Life, Liberty and the Defense of Dignity: The Challenge for Bioethics* (San Francisco, CA: Encounter Books, 2004), pp. 288–97.
12. Wendell Berry, *The Unsettling of America: Culture and Agriculture* (New York, NY: Avon Books, 1978), p. 62.
13. Pope Benedict XVI, *In the Beginning: A Catholic Understanding of the Story of the Creation and the Fall* (Grand Rapids, MI: Eerdmans, 1995), p. 34. See also LS 116.
14. Pope Benedict XVI, *In the Beginning*, p. 37.
15. Michael Hanby, "Gospel of Creation and the Technocratic Paradigm: Reflections on a Central Teaching of Laudato Si'," *Communio: International Catholic Review*, 42 (4) (2015), 733.
16. See, for example, W. David Montgomery, "The Flawed Economics of *Laudato Si'*," *The New Atlantis: A Journal of Technology and Society*, 47 (Fall 2015), 31–44.
17. On the limiting effects of economic methodology on a more adequate understanding or rationality, see Andrew Yuengert, *Approximating Prudence: Aristotelian Practical Wisdom and Economic Models of Choice* (New York, NY: Palgrave Macmillan, 2012), esp. pp. 157–76.
18. See LS 49, and also Frank Pasquale, "A Critique of Mastery and an Ethics of Attunement: From *Spe Salvi* to *Laudato Si'*" in Chapter 8 of the present volume.
19. Wendell Berry, "The Total Economy" in Doug Bandow and David L. Schindler (eds.), *Wealth, Poverty and Human Destiny* (Wilmington, DE: ISI Books, 2003), p. 419.
20. For a more extended discussion of the importance of local economies in *Laudato Si'*, particularly for countering the tendency to technocratic abstraction in economic thinking, see Patrick Fleming, "Economics, Ecology and our Common Home: The Limits of a Preference-based Approach to Human Behavior," *Communio: International Catholic Review*, 42 (4) (2015), esp. 715–23.

21. Compare Marcel Gauchet: "Modern man's relation to nature, which is both possessive and transformative, is inseparable from individualism as the organizing vision of the human-social phenomenon": *The Disenchantment of the World: A Political History of Religion*, trans. Oscar Burge (Princeton, NJ: Princeton University Press, 1997), p. 94.
22. See "Dialogo del Santo Padre con i Vescovi della Polonia" (Kraków, 27 July 2016): http://press.vatican.va/content/salastampa/it/bollettino/pubblico/2016/08/02/0568/01265.html#en
23. For an expansion on this topic, see Hanby, "Gospel of Creation."
24. According to one analysis, at least 6,000 male-to-female sex reassignment surgeries alone were performed in the United States in the 1970s, at least 9,000 in the 1980s, and at least 14,000 from 1990 to 2002: Lynn Conway, "How Frequently does Transsexualism Occur?": http://ai.eecs.umich.edu/people/conway/TS/TSprevalence.html. Yet, even in 2009, a study in the major professional journal for reconstructive surgery deemed that "the magnitude of benefit and harm for individual surgical procedures cannot be estimated accurately using the current available evidence" ("Evaluation of Surgical Procedures for Sex Reassignment: A Systematic Review," *Journal of Plastic, Reconstructive and Aesthetic Surgery*, 62.3 (2009), 294–306); and, in 2012, the American Psychiatric Association's task force on gender identity disorder concluded that "quality of evidence pertaining to most aspects of treatment in all subgroups was determined to be low." ("Report of the American Psychiatric Association Task Force on Treatment of Gender Identity Disorder," Archives of Sexual Behavior, 41 (4) (2012), 759–96).
25. Cardinal Jorge Mario Bergoglio, "Same-sex Marriage: The Letter from Cardinal Jorge Mario Bergoglio, S.J. to the Carmelite Nuns of Buenos Aires" (June 22, 2010): www.catholichawaii.org/media/224245/bergoglio_to_carmelite_sisters.pdf
26. Nico Lang, "Pope Francis' LGBT failure: Vulnerable Transgender Community Needs His Active Support Now," *Salon*, October 6, 2016: www.salon.com/2016/10/06/pope-francis-lgbt-failure-vulnerable-transgender-community-needs-his-active-support-now/
27. Cardinal Bergoglio, "Same-sex Marriage."
28. Francis DeBernardo, "Some Hope But Not Much Joy for LGBT Catholics in Pope's 'Joy of Love' Document" (New Ways Ministry, April 8, 2016): https://newwaysministryblog.wordpress.com/2016/04/08/popes-joy-of-love-is-not-joyful-for-lgbt-catholics-2/
29. "Cardinal Turkson Urges Bishops to Spread Message of Laudato Si'," *Zenit*, January 26, 2016: https://zenit.org/articles/cardinal-turkson-urges-bishops-to-spread-message-of-laudato-si/

6

Towards a Bright Mountain: *Laudato Si'* and the Critique of Technology

Zachary Loeb

We hate the people who make us form the connections we do not want to form.

– Simone Weil[1]

1 REPAIRING OUR COMMON HOME

When confronted with the unsettling realities of the world it is easy to feel overwhelmed and insignificant. This feeling of powerlessness may give rise to a temptation to retreat – or simply to shrug. Some may suspect that they bear responsibility for the current state of affairs, yet the scale of the problems makes individuals doubtful that they can make a difference. In this context, the refrain "well, it could always be worse" becomes a sort of inured coping strategy, though this dark prophecy has a tendency to prove itself true, week after week and year after year. After all, just saying that things could be worse does nothing to prevent those things from further deterioration. Alas, it can be rather liberating to decide that one is powerless, to conclude that one's actions do not matter, to imagine that one will be long dead by the time the bill comes due – for taking such positions enables one to avoid doing something difficult: changing.[2]

And yet, a change is coming. Indeed, the change is already here. The question is whether people are willing to consciously change to meet this challenge or if they will only change when they truly have no other option.

The matter of change is a core feature of Pope Francis's *Encyclical on Climate Change and Inequality: On Care for Our Common Home* – also known as *Laudato Si'* ("Praise Be to You").[3] The full title of the encyclical makes it easy to focus on the environmental aspect of Pope Francis's text, but it is reductive to cast *Laudato Si'* as simply being about the environment. Framing the encyclical in such a simplistic way makes it appear quaint – and may lead to many concluding that they do not need to spend the time genuinely wrestling with each of the encyclical's 245 sections. True, Pope Francis is interested in climate change, but in the encyclical he proves just as interested in the changes in the social, economic, and political climate that have allowed climate change to advance. The importance of *Laudato Si'* is precisely that it is less about climate change than it is about the need for humanity to change, for, "we cannot adequately combat environmental degradation unless we attend to causes related to human and social degradation" (LS 48).

And though the encyclical is filled with pithy, highly quotable aphorisms, truly engaging with the text requires us to consider it in full.

Lest there be any doubt, *Laudato Si'* is a difficult text to read – not because it is written in archaic prose, or because it assumes that the reader is learned in theology, but because it is discomforting. *Laudato Si'* does not tell the reader that they are now responsible for the world; instead it reminds them that they have always been responsible for the world. The encyclical calls on those with their heads in the clouds (or head in "the cloud") to see the mess in which they are standing. However, Pope Francis refuses to take the reassuring route of claiming that there is an easy solution. What *Laudato Si'* represents is an unabashed ethical assault on high-tech/high-consumption life in affluent nations. Yet it is not an angry diatribe or a woebegone dirge. Insofar as the encyclical represents a hammer, it is not as a blunt instrument with which one bludgeons foes, but instead a useful tool that one might take up to pull out the rusted old nails in order to build again. After all, "Humanity still has the ability to work together in building our common home" (LS 13).

Laudato Si' is a work of intense, even radical, social criticism in the fine raiment of a papal encyclical. The text contains a cutting critique of technology, an ethically rooted castigation of capitalism, an impassioned defense of the environment, and a demand that people accept responsibility for their actions. There is much in *Laudato Si'* that those well versed in activism, organizing, environmentalism, critical theory, the critique of technology, radical political economy (and so forth) will find familiar – and it is a document that has much to contribute to those aforementioned discourses. While the encyclical contains numerous references to Jesus, God, the Church, and the saints – it is clear that Pope Francis intends the document for a wide readership. Indeed, those versed in religious traditions other than Catholicism will likely find much in the encyclical that echoes their own beliefs – and the same can likely be said of many who do not subscribe to a particular religion. While *Laudato Si'* speaks, at times, to the religious obligation of believers, Pope Francis makes a point of being inclusive to those of different, or no, faith – an inclusion that speaks to his recognition that the problems facing humanity can only be solved by humanity. Insofar as "we need only take a frank look at the facts to see that our common home is falling into serious disrepair" (LS 61) the terms "common home" and "we" make it clear that this is a matter affecting the planet and all those – regardless of their faith – who dwell there.

Nevertheless, some sections in *Laudato Si'* may seem less inclusive than others. Pope Francis stands firm in his commitment to the poor, and makes frequent comments about the rights of indigenous communities – but he does not have particularly much to say about women. The importance of equality is a theme that is revisited numerous times in *Laudato Si'* and, though Pope Francis addresses his readers as "sisters and brothers," it is worth wondering whether this entails true equality between all people, regardless of gender. Pope Francis laments the lack of concern being shown to those already struggling against the rising tide of climate change, noting: "Our lack of response to these tragedies involving our brothers and sisters points to the loss of that sense of responsibility to our fellow men and women upon which all civil society is founded" (LS 25). Yet it is worth pushing on this "sense of responsibility" – and doing so should entail a recognition that, too often, "civil society" has been founded on an emphasis on "fellow men" and not necessarily upon women. A belief that "all men are created equal" can easily exclude women, and many men as well. However, Pope Francis does not merely allude to "civil society"; rather, he evokes the more ethically tinged notion of "responsibility." This mixture of "civil society" and "responsibility" is reminiscent of the idea of "obligation" that Simone Weil developed in *The Need for Roots*. To Weil, "There exists an obligation towards every human being for the sole reason that he or she is a human being, without any other condition requiring to be fulfilled, and even without any recognition of such obligation on the part of the individual concerned."[4] To recognize that the obligation is due to "the human being as such" – which seems to be something Pope Francis is claiming – necessitates acknowledging that, unless one is careful, "the human being" may still be defined as male. And this is a bias that can easily be replicated, even in encyclicals that tout the importance of equality. While such aspects may give some readers cause to furrow their brows, it would be unfortunate if such shortcomings led people to dismiss the encyclical completely.

Laudato Si' is not a document that one really reads; rather, it is a text with which one wrestles. And, as befits a piece written by a former nightclub bouncer, *Laudato Si'* proves to be a challenging combatant. However, the easiest way to emerge victorious from a bout is to refuse to engage in it in the first place – a tactic that was manifested in every comment that "the Pope should stick to religion." Granted, *Laudato Si'* does not claim to be a scientific document, and Pope Francis does not claim to be a scientist; yet he accepts the scientific consensus on climate change (LS 23). Pope Francis casts climate change as "a global problem with grave implications" the effects of which will be felt widely, though the most severe impacts will not be distributed evenly – and thus climate change "represents one of the principal challenges facing humanity" (LS 25).

As has been previously noted, *Laudato Si'* is as much a work of stinging social criticism as it is a theological document. It is a text that will be read differently when it is filtered through different people – be they workers, theologians, activists, scholars, and the list could go on – with knowledge of the various topics upon which the encyclical touches. And yet, one of the most striking aspects of the encyclical – that which poses a particular challenge to the status quo – is the way in which the document engages with technology.

For, it may well be that *Laudato Si'* will change the tone of current discussions around technology and its role in the contemporary world – at least, one might hope that it will do so.

II MEET THE NEW GODS, NOT THE SAME AS THE OLD GOD

Perhaps being a person of faith makes it easier to recognize the religious faith of others. Or, put another way, perhaps belief in God makes one attuned to the appearance of new gods. While a recent Pew study has suggested that, in recent years, the number of individuals who do not adhere to a particular religious doctrine has risen,[5] *Laudato Si'* suggests – though not in these terms – that people may have simply turned to new religions. In *To Be and to Have*, Erich Fromm uses the term "religion" not to denote a belief system reliant on God, "but to *any group-shared system of thought and action that offers the individual a frame of orientation and an object of devotion*."[6]

Though the author of *Laudato Si'*, obviously, ascribes to a belief system with a place for God – the encyclical is staked out in opposition to the rise of a "group-shared system of thought" that has come to offer many people both "a frame of orientation and an object of devotion." Pope Francis warns his readers against giving fealty and adoration to false gods – gods that are as appealing to atheists as they are to old-time-believers. And while *Laudato Si'* is not a document that seeks to draw people into the Catholic Church, it is a document that warns people against treating technology as a religion. If an apple led to expulsion from the Garden of Eden, then it is doubtful that we can return there by biting into an Apple product.

It is worth recognizing, that there are many reasons why the religion of technology so easily wins converts. As the encyclical acknowledges, the world is a mess and the news reports are filled with a steady flow of horrors. Yet, one of the few areas that continually operates as a site of unbounded optimism is the missives fired off by the technology sector and its boosters. Wearable technology, self-driving cars, the Internet of Things, delivery drones, artificial intelligence, virtual reality – technology provides a vision of the future that is not glumly fixated on rising sea levels. Indeed, some techno-enthusiasts are even predicting that, through the power of techno-science, humans may not be far off from being able to bring back species that had previously become extinct.[7]

Technology has become a site of millions of minor miracles that have drawn legions of adherents to the technological god and its sainted corporations – the high-technology of today seems increasingly to be presented in a way that encourages people to bask hopefully in its uncanny glow. Contemporary technology – especially of the Internet-connected variety – promises individuals that they will never be alone, that they will never be bored, that they will never get lost, and that they will never have a question for which they cannot execute a web search and find an answer. And when people suddenly find themselves without their device, they feel alone – or as Katherine Hayles put it, as though "her hand had been amputated."[8] If some religions spoke of a god who was always watching, and always with the believer, than the smartphone replicates and reifies these beliefs – for it is always with the believer, and it is always watching. To return to Fromm's description of religion, it should be fairly apparent that technology today provides people with "a frame of orientation and an object of devotion." It is thus not simply that technology comes to be presented as a solution to present problems, but that it comes to be presented as a form of salvation. Why pray if "there's an app for that"? In *Laudato Si'*, Pope Francis warns against this new religion, observing that, "Life gradually becomes a surrender to situations conditioned by technology, itself viewed as the principle key to the meaning of existence" (LS 110).

This leads to a question as to what "the meaning of existence" supplied by contemporary technology looks like. The various denominations of the religion of technology are skilled at offering appealing answers to this question, filled with carefully tested slogans about making the world "more open and connected" or signaling that their users "think different." Such slogans are more appealing than the world of vast economic inequality, rising sea levels, misery, existential uncertainty, mountains of filth discarded by affluent nations (which includes e-waste), and grim prospects described in *Laudato Si'*. By comparison, the religion of technology provides a shiny vision of the future, with the promise of escape from earthly concerns through virtual reality, delivery on demand, and the truly transcendent dream of becoming one with machines. The religion of technology is not concerned with the next life, or with the lives of future generations; it is about constructing a new Eden in the now, for those who can afford the right toys – even if constructing such a heaven consigns much of the world's population to hell. People may not be bending their necks in prayer, but they're certainly bending their necks to glance at their smartphones.

This transformation, whereby technology comes to be treated as an object of worship, is not new. In the estimation of the historian David Noble, in his aptly titled book *The Religion of Technology*, "the expectation of ultimate salvation through technology, whatever the immediate human and social costs, has become the unspoken orthodoxy, reinforced by a market-induced enthusiasm for novelty and sanctioned by millenarian yearnings for new beginnings."[9] This was also a matter of great concern, and much rumination, for Jacques Ellul, whose expansive scholarly output includes numerous books on the social impact of technology as well as many texts focused on theology. In Ellul's estimation, "the fact is that technology is felt by modern man as a sacred phenomenon."[10] And lest Ellul be accused of being unclear, he puts it even more starkly, noting that the social perception becomes one wherein "Technology is the god who saves."[11] And Lewis Mumford would deride such sentiments as "the ultimate religion of our seemingly rational age – The Myth of the Machine!"[12]

Against this religious embrace of technology, and the elevation of its evangels, *Laudato Si'* puts forth a reminder that one can, and should, appreciate the tools that have been invented – but that one should not worship them. As Erich Fromm put it, "The question is not one of *religion or not?* but of *which kind of religion?*" and as he continues "*we are what we are devoted to, and what we are devoted to is what motivates our conduct.*"[13] Technology is not the only false religion against which the encyclical warns – the cult of self-worship, unbridled capitalism, the glorification of violence, and the consumerism are all considered as false faiths. They draw adherents in by proffering salvation and prescribing a simple course of action – but instead of allowing their faithful true transcendence they instead warp their followers into sycophants who, as Ellul put it, "fly into a mystic rage when one contests the validity of technology, and from that point on they call down doom on the contesting person."[14]

Yet the particularly nefarious aspect of the religion of technology is the way in which it is a faith to which many subscribe without their necessarily being aware of it. This is particularly significant in the way that it links to the encyclical's larger concerns. Those in affluent nations who enjoy the "goods" of high-tech lifestyles are spared many of the serious downsides of high technology. Certainly, individuals may complain of aching necks, sore thumbs, difficulty sleeping, and a creeping sense of dissatisfaction – but such issues do not tell of the true cost of technology. Those who are "alone together," to borrow Sherry Turkle's term, are often not the ones who are "together" with a mountain of slowly degrading e-waste.[15] What often goes unseen by those enjoying their smartphones are the regimes of mineral extraction, the labor conditions of high-tech device assembly, and the toxic wreckage of e-waste dumps. Furthermore, insofar as high-tech devices require large amounts of energy, it is worth considering the degree to which high-tech lifestyles contribute to the voracious energy consumption that helps to drive climate change. As a case in point, Pope Francis alludes to the "increasing use and power of air-conditioning" as a sign of a harmful habit of consumption – but "the cloud" requires a great deal of air conditioning to keep cool (LS 55). The benefits of high-tech devices come swiftly to their users who are generally insulated from the "slow violence" (to use Rob Nixon's term) that such devices ultimately inflict.[16]

And though *Laudato Si'* may have a particular view of salvation, one need not be Catholic to understand Pope Francis's message that salvation will not come through worshipping technology.

III *LAUDATO SI'* AS A CRITIQUE OF TECHNOLOGY

Relatively early in the encyclical, Pope Francis decries how, against the background of "media and the digital world . . . the great sages of the past run the risk of going unheard amid the noise and distractions of an information overload" (LS 47). A glance at the citations in *Laudato Si'* makes it fairly apparent who Pope Francis considers many of these "great sages" to be. For the most part, Pope Francis cites the encyclicals of his predecessors, declarations from Bishops' conferences, the bible, and theologians who are safely ensconced in the Church's wheelhouse. While such citations certainly help to establish that the ideas being put forth in *Laudato Si'* have been circulating in the Catholic Church for some time, Pope Francis's invocation of "great sages of the past" raises a larger question. How much of the encyclical is a reiteration of older ideas that have gone "unheard"? In fairness, the social critique being advanced by *Laudato Si'* may strike many people as novel – particularly in terms of its ethically combative willingness to take on technology – but it may be that the significant thing about *Laudato Si'* is not that the message is fresh, but that the messenger is new. Without wanting to decry or denigrate *Laudato Si'* it is worth noting that much of the argument being presented in the document could previously be found in works by thinkers associated with the critique of technology, notably Lewis Mumford and Jacques Ellul – though those familiar with the work of the likes of Neil Postman, Günther Anders, Hans Jonas, Albert Borgmann, and Langdon Winner (to name but a few) will also see many parallels to their work. Indeed, the following statement, from Lewis Mumford's *Art and Technics*, could have appeared in *Laudato Si'* without seeming out of place: "the machine has become our main source of magic, and it has given us a false sense of possessing godlike powers. An age that has devaluated all its symbols has turned the machine itself into a universal symbol: a god to be worshipped."[17]

The critique of technology does not represent a cohesive school of thought – rather, it is a tendency that stretches across several fields that places particular emphasis on skepticism towards the promises of technology. Within the critique of technology, it is not a matter of anything so ridiculously reductive as "technology is bad" but of considering what alternative forms technology could take, be they: "democratic technics,"[18] "convivial tools,"[19] "appropriate technology,"[20] "liberatory technology,"[21] and so forth. Yet what is particularly important is the fact that the critique of technology is directly tied to a critique of the broader society.

With the rise of ever more powerful and potentially catastrophic technological systems, many thinkers associated with the critique of technology began issuing dire warnings about the techno-science-wrought danger in which humanity had placed itself. With the appearance of the atomic bomb, humanity had invented the way to potentially bring an end to the whole of the human project. Galled by the way in which technology seemed to be drawing ever more power to itself, Ellul warned of the ascendance of "technique" while Mumford cautioned of the emergence of "the megamachine" with such terms being used to denote not simply technology and machinery but the fusion of techno-science with social, economic, and political power.[22] In *Laudato Si'*, Pope Francis does not use Mumford's or Ellul's terms but the way in which he writes about the "technological paradigm" or "technocratic paradigm" comfortably matches what Ellul and Mumford had sought to convey.

When Pope Francis writes, "The technological paradigm has become so dominant that it would be difficult to do without its resources and even more difficult to utilize them without being dominated by their internal logic" (LS 108); or, "the new power structures based on the techno-economic paradigm may overwhelm not only our politics but also freedom and justice" (LS 53); or "The alliance between the economy and technology ends up sidelining anything unrelated to its immediate interests" (LS 54); these are comments that are squarely in line with Ellul's comment that: "*Technical civilization* means that our civilization is constructed *by* technique (makes a part of civilization only that which belongs to technique), *for* technique (in that everything in this civilization must serve a technical end), and *is* exclusively technique (in that it excludes whatever is not technique or reduces it to technical forms)."[23]

A particular sign of the growing dominance of technology, and the techno-utopian thinking that everywhere evangelizes for it, is the belief that for every problem there is a technological solution. Such wishful thinking about technology as the universal panacea was a tendency highly criticized by thinkers such as Mumford and Ellul. Pope Francis chastises the prevalence of this "blind confidence in technical solutions" (LS 14) and he decries "Technology, which, linked to business interests, is presented as the only way of solving these problems" (LS 20). There is more than a passing similarity between these lines from the encyclical and the following quotation from Lewis Mumford's book *Technics and Civilization*: "But the belief that the social dilemmas created by the machine can be solved merely by inventing more machines is today a sign of half-baked thinking which verges close to quackery."[24]

The purpose of juxtaposing these comments from Pope Francis with lines from Mumford and Ellul is not to undermine the significance of *Laudato Si'* but to demonstrate that the sentiments regarding technology that Pope Francis seeks to convey are very similar to ideas voiced by prominent critics of technology. If one wanted to do so it would not be an overly difficult task to comb through *Laudato Si'* – particularly "Chapter Three: The Human Roots of the Ecological Crisis" – and find a couple of paragraphs by Mumford, Ellul, or another prominent critic of technology in which precisely the same thing is being said. While, as has been previously noted, the critique of technology does not refer to a single coherent body of thought, if one wanted to distill the essence of this not quite official sub-field into two sentences one could do significantly worse than the following lines from *Laudato Si'*: "We have to accept that technological products are not neutral, for they create a framework which ends up conditioning lifestyles and shaping social possibilities along the lines dictated by the interests of certain powerful groups. Decisions which may seem purely instrumental are in reality decisions about the kind of society we want to build" (LS 107). The line "technological products are not neutral" may come as something of a disquieting statement to some, but it has long been understood in scholarship around technology's impacts.[25]

Nevertheless, it is the emphasis placed on the matter of "the kind of society we want to build" that is of particular importance. For the encyclical does not simply lament the state of the technological world; it advances an alternative vision of technology – one that recognizes the tremendous potential of technological advances but sees how this potential repeatedly goes awry. *Laudato Si'* is a document that is skeptical of the belief that smartphones have made people happier, and it is a text that shows a clear unwillingness to believe that large tech companies are driven by much other than their own interests. The encyclical bears the mark of a writer who believes in a powerful God and that deity's prophets, but has little time for would-be all-powerful corporations and their lust for profits. One of the themes that ran continuously throughout Mumford's work was his belief that the "good life" had been overshadowed by the pursuit of the "goods life"[26] – and a similar theme runs through *Laudato Si'* wherein the analysis of climate change, the environment, and what is owed to the poor, is couched in a call to reinvigorate the "good life" while recognizing that for many people the "goods life" has become a substitute for it.

Despite the power of the "technological paradigm," Pope Francis remains hopeful regarding the power of people, writing: "We have the freedom needed to limit and direct technology; we can put it at the service of another type of progress, one which is healthier, more human, more social, more integral" (LS 112). Here, Pope Francis is arguing for the need for, to use Mumford's terminology, "democratic technics" to replace "authoritarian technics."[27] One thing that importantly sets Pope Francis apart from critics such as Mumford and Ellul is that he tends to be rather optimistic about the potential that individuals have for limiting and directing technology. A fear shared among numerous critics of technology was the belief that the concentration of power in "technique" or "the megamachine" or the "technological paradigm" gradually eliminated the freedom to limit or direct it. It was not that critics did not believe in alternatives, but that they feared that such paths were quickly incorporated back into the "technological paradigm." As Ellul observed: "to be in technical equilibrium, man cannot live by any but the technical reality ... the more his needs are accounted for, the more he is integrated into the technical matrix."[28] In other words, "technique" gradually eliminates the alternatives to itself. To live in a society shaped by such forces requires an individual to submit to those forces as well.

What *Laudato Si'* claims, on the contrary, is that it is not too late, and that people still have the ability "to limit and direct technology" provided that they tear themselves away from their high-tech hallucinations.

The focus on technological matters is one of the features that sets *Laudato Si'* apart from other texts decrying what industrial capitalism has done to the planet. For what the encyclical does so impressively is that it expressly links environmental destruction and the neglect of the poor with the allegiance to high-tech devices. On a regular basis, there appe ar numerous books and articles that express discontent with the current state of the technological world – and yet, too often, the authors of such texts seem terrified of being branded as "anti-technology." Therefore, the authors tie themselves in knots trying not to evangelize for technology but at the same time refusing to truly challenge that belief system – and as a result they easily become the permitted voices of dissent. They suggest an alternative tune for the hymns instead of recognizing that it is a faith premised upon worshipping a false god. After all, one is permitted to say that Google is getting too big, that the Apple Watch is unnecessary, and that the Internet has unleashed a surveillance nightmare – but to say: "There is a growing awareness that scientific and technological progress cannot be equated with the progress of humanity and history, a growing sense that the way to a better future lies elsewhere" (LS 113). Now that is a radical sentiment!

Such was the type of argument being made by the likes of Mumford and Ellul – and numerous other critics of technology whose works also deserve to be considered. Alas, the present predicament makes it woefully evident that too few heeded their warnings. Indeed, a *leitmotif* that can be detected among the works of many critics of technology is an earnest desire to be wrong, as Mumford wryly remarked: "I would die happy if I knew that on my tombstone could be written these words, 'This man was an absolute fool. None of the disastrous things that he reluctantly predicted ever came to pass!'"[29] Yet Mumford was not an "absolute fool"; he was acutely prescient. The likes of Mumford and Ellul easily number among the ranks of "the great sages of the past" who have largely gone unheard, though it may be the case that such voices are not simply "unheard" but deliberately silenced.

Laudato Si' commits a profound act of heresy against the contemporary religion of technology – it dares to suggest that we have fallen for the PR spin about the devices in our pockets, asks us to consider if these devices are truly filling an existential void, and reminds us that we need not be passive consumers of technology. These arguments about technology are not new, and it is not new to make them in ethically rich or religiously loaded language; however, these are arguments that are *verboten* in contemporary discourse about technology. Those who make such claims are regularly derided as "Luddites" or "NIMBYs" and treated as worthy of mockery.[30] And yet the historic Luddites were simply workers who felt that they had the freedom "to limit and direct technology,"[31] and as anybody who knows about e-waste can attest when people in affluent nations say "Not in My Back Yard" the toxic refuse simply winds up in somebody else's back yard.[32]

Pope Francis writes that: "It has become countercultural to choose a lifestyle whose goals are even partly independent of technology, of its costs and its power to globalize and make us all the same" (LS 108). And thus, what *Laudato Si'* may represent is an important turning point in discussions around technology, and a vital opportunity for a serious critique of technology to reemerge. For what *Laudato Si'* does is advocate for a new cultural paradigm based upon harnessing technology as a tool instead of as an absolute. Furthermore, the inclusion of such a powerful critique of technology in a widely discussed (and, it is to be hoped, widely read) encyclical represents a point at which rigorously critiquing technology may be able to become less "countercultural." It is a rare text that has the bravery to state: "you do not need that."

Like many works associated with the critique of technology, *Laudato Si'* is a fierce piece of social criticism that recognizes that one cannot truly claim to critique a society without being willing to turn an equally critical gaze towards the way in which society creates and uses technology. The critique of technology is not new, but it has been sorely underrepresented in contemporary debates around technology. It has been cast as the province of outdated doom mongers, but, as Pope Francis demonstrates, such a critique remains vital.

Too often of late, discussions about technology are conducted through rose-colored glasses – or worse, virtual reality headsets. *Laudato Si'* dares actually to look at technology – and to demand that others do the same.

IV THE BRIGHT MOUNTAIN

Doom is easy: all it requires of us is that we do nothing. And what can be simpler than doing nothing? Besides, popular culture has made us quite comfortable with the images of post-apocalyptic deathscapes and collapsing cities. And yet the question to ask of every piece of dystopian fiction is: "What did the world that paved the way for this terrible one look like?" To which the follow-up question should be: "Did it look just like ours?" And to this, yet another follow-up question needs to be posed: "Why didn't people do something?"

Prior to penning the introduction to the Melville House edition of the encyclical, Naomi Oreskes (along with Erik Conway) wrote a short book bearing the uplifting title *The Collapse of Western Civilization* in which Oreskes and Conway analyze present inaction as if from the future, noting that: "the people of Western civilization knew what was happening to them but were unable to stop it. Indeed, the most startling aspect of this story is just how much these people knew, and how unable they were to act upon what they knew."[33] This speaks to the fatalistic belief that, despite what we know, things are not going to change, or that if change comes it will already be too late. It is hard not to wince when Oreskes and Conway characterize 2009 as "the last best chance" to act.[34]

This matter, of continually flouted warnings, was disturbingly ruminated upon in a slim volume entitled *Uncivilisation: The Dark Mountain Manifesto*. It is the foundational text of a group of writers, artists, activists, and others that takes seriously the notion that we are not going to change in time. As the manifesto's authors write: "Secretly, we all think we are doomed: even the politicians think this; even the environmentalists. Some of us deal with it by going shopping. Some deal with it by hoping it is true. Some give up in despair. Some work frantically to try and fend off the coming storm."[35] But the point is that change is coming – whether we believe it or not, and whether we want it or not.

The easy retort to the claim "secretly, we all think we are doomed" is "I do not think we are doomed, secretly or openly" – though it may be that such a dismissal is just proof of the continuing veracity of Günther Anders's observation, written regarding the danger of nuclear weapons, that "it is our capacity to fear which is too small and which does not correspond to the magnitude of today's danger."[36] After all, to read climatologists' predictions and then to watch politicians grouse, while mining companies seek to extract even more fossil fuels, is to hear that "secret" voice grow louder. Indeed, in the time since the encyclical's initial publication and this writing, scientists' warnings have continued – with some already suggesting that the goals of the Paris accords were too modest.[37] Apocalyptic predictions are nothing new, but here we still are, and so many people simply shrug, while others worry that putting too much emphasis on woebegone premonitions and "catastrophism" overwhelms people and leaves them unable and unwilling to act.[38] Perhaps this is why Al Gore's film *An Inconvenient Truth* concludes by telling people not that they must be willing fundamentally to alter their high-tech/high-consumption lifestyles but instead that they should recycle more often.

So what is one to do? The desire to retreat from the cacophony of modern society is nothing new, and can easily sow the fields in which reactionary ideologies can grow. Particularly problematic is that the rejection of the modern world often entails a sleight of hand whereby those in affluent nations are able to shirk their responsibility to the world's poor even as they walk somberly, flagellating themselves into the foothills. Apocalyptic romanticism, whether it be of the accelerationist or primitivist variety, paints an evocative image of the world of today collapsing so that a new world can emerge – but what *Laudato Si'* counters with is a morally impassioned cry to think of the billions of people who will suffer and die. Some may dream of building a better world on the ruins of this one, but a lot of bodies will be buried beneath those ruins. *Laudato Si'* counters with an exhortation to think of those for whom fleeing to the foothills is not an option.

"Doomsday predictions can no longer be met with irony or disdain. We may well be leaving to coming generations debris, desolation and filth" (LS 161). These lines from Pope Francis could be the twenty-first-century replacement for "keep calm and carry on." For what *Laudato Si'* makes clear is that now is not the time to "keep calm" but to get very busy, and it is a text that knows that if we "carry on" than we are skipping aimlessly towards the cliff's edge. And yet one of the elements of the encyclical that needs to be highlighted is that it is a document that does not look hopefully towards a coming apocalypse. In the encyclical, environmental collapse is not seen as evidence that biblical preconditions for Armageddon are being fulfilled. The sorry state of the planet is not the result of God's plan but is instead the result of humanity's inability to plan. The problem is not evil, for, as Simone Weil wrote: "It is not good which evil violates, for good is inviolate: only a degraded good can be violated."[39] It is that the good of which people are capable is rarely the good that they achieve. Even as possible tools for building the good life – such as technology – are degraded and mistaken for the good life. And thus the good is wasted, though it has not been destroyed.

Throughout *Laudato Si'*, Pope Francis praises the merits of an ascetic life. And though the encyclical features numerous references to Saint Francis of Assisi, the argument is not that we must all abandon our homes to seek out sanctuary in the woods; instead, the need is to learn from the sense of love and wonder with which Saint Francis approached nature. Complete withdrawal is not an option; to do so would be to shirk our responsibility – we live in this world and we bear responsibility for it and for other people. We do not need to take up residence in the woods like latter-day acolytes of Francis of Assisi; rather, we need to take that spirit and live it wherever we find ourselves. In the encyclical's estimation, those living in affluent nations cannot seek to slip quietly from the scene, nor can they claim that they are doing enough by bringing their own bags to the grocery store. Responsibility entails recognizing that the lifestyles of affluent nations have helped to sow misery in many parts of the world – and that it is unethical for us to try to save our own cities without realizing the part that we have played in ruining the cities of others.

As Pope Francis writes:

> "Many things have to change course, but it is we human beings above all who need to change. We lack an awareness of our common origin, of our mutual belonging, and of a future to be shared with everyone. This basic awareness would enable the development of new conviction, attitudes and forms of life. A great cultural, spiritual and educational challenge stands before us, and it will demand that we set out on the long path of renewal." (LS 202)

Laudato Si' does not suggest that we can escape from our problems, that we can withdraw, or that we can "keep calm and carry on." And though the encyclical is not a manifesto, if it were one it could possibly be called "The Bright Mountain Manifesto." For what *Laudato Si'* reminds its readers time and time again is that, even though we face great challenges, it remains within our power to address them, though we must act soon and decisively if we are to effect a change. We do not need to wander towards a mystery-shrouded mountain in the distance, but work to make the peaks near us glisten – it is not a matter of retreating from the world but of rebuilding it in a way that provides for all. Nobody needs to go hungry, our cities can be beautiful, our lifestyles can be fulfilling, our tools can be made to serve us as opposed to our being made to serve tools, people can recognize the immense debt they owe to each other – and, working together, we can make this a better world.

Doing so will be difficult and will require significant changes. But *Laudato Si'* is a document that believes that people can still accomplish this. The encyclical warns against placing faith in technology. Yet, in the end, *Laudato Si'* is less about having faith in God than it is about having faith in people.

Notes

1. Simone Weil, *Gravity and Grace* (London: Routledge Classics, 2002), p. 139.
2. An earlier version of this chapter was originally published in the Digital Studies section of Boundary 2, under the title "Towards a Bright Mountain: *Laudato Si'* as Critique of Technology" (July 1, 2015). It received editing before posting there from David Golumbia. The original version is still accessible at: www.boundary2.org/2015/07/towards-a-bright-mountain-laudato-si-as-critique-of-technology/
3. Pope Francis, *Encyclical Letter Laudato Si' of the Holy Father Francis on Care for Our Common Home* (Vatican Press, 2015).
4. Simone Weil, *The Need for Roots* (London: Routledge Classics, 2002), p. 5.
5. Michael Lipka, "A closer look at America's rapidly growing religious 'nones'," *Pew Research Center*, May 13, 2015: www.pewresearch.org/fact-tank/2015/05/13/a-closer-look-at-americas-rapidly-growing-religious-nones/
6. Erich Fromm, *To Have or to Be* (New York, NY: Harper & Row, Publishers, 1976), p. 135 (italics in original).
7. Stewart Brand, "Rethinking Extinction," *Aeon*, April 21, 2015: http://aeon.co/magazine/science/why-extinction-is-not-the-problem/
8. N. Katherine Hayles, *How We Think: Digital Media and Contemporary Technogenesis* (Chicago, IL: University of Chicago Press, 2012), p. 2.
9. David Noble, *The Religion of Technology* (New York, NY: Penguin, 1999), p. 207.
10. Jacques Ellul, *The New Demons* (New York, NY: The Seabury Press, 1975), p. 71.
11. Ibid., p. 73.
12. Lewis Mumford, *Findings and Keepings: Analects for an Autobiography* (New York, NY: Harcourt, Brace and Jovanovich, 1975), p. 375.
13. Fromm, *To Have or to Be*, pp. 135–36 (italics in original text).
14. Ellul, *The New Demons*, p. 73.
15. Sherry Turkle, *Alone Together: Why We Expect More from Technology and Less from Each Other* (New York, NY: Basic Books, 2012).
16. Rob Nixon, *Slow Violence and the Environmentalism of the Poor* (Cambridge, MA: Harvard University Press, 2010). Nixon defines "slow violence" as "a violence that occurs gradually and out of sight, a violence of delayed destruction that is dispersed across time and space, an attritional violence that is typically not viewed as violence at all." (ibid., p. 2).
17. Lewis Mumford, *Art and Technics* (New York, NY: Columbia University Press, 2000), p. 138.
18. Lewis Mumford, "Authoritarian and Democratic Technics," Technology and Culture, 5(1) (1964), 1–8.
19. Ivan Illich, Tools for Conviviality (New York, NY: Harper & Row Publishers, 1973).
20. E. F. Schumacher, *Small is Beautiful: Economics as if People Mattered* (New York, NY: Harper Perennial, 2010).
21. Murray Bookchin, "Towards a Liberatory Technology" in Post-Scarcity Anarchism (Oakland: AK Press, 2004), pp. 41–84.
22. Jacques Ellul, *The Technological Society* (New York, NY: Vintage Books, 1964); *The Myth of the Machine: The Pentagon of Power* (New York, NY: Harcourt, Brace and Jovanovich, Inc., 1970).
23. Ellul, *The Technological Society*, p. 128.

24. Lewis Mumford, *Technics and Civilization* (Chicago, IL: University of Chicago Press, 2010), p. 367.
25. See Langdon Winner, *The Whale and the Reactor: A Search for Limits in an Age of High Technology* (Chicago, IL: The University of Chicago Press, 1989).
26. This is a theme to which Mumford returned repeatedly throughout his work. In the bibliography for *The Pentagon of Power* he referred to his earlier book *The Story of Utopias* as having been "My first exposition of the fundamental difference between the good life and the 'goods life." (Lewis Mumford, *The Myth of the Machine: The Pentagon of Power* (New York, NY: Harcourt, Brace and Jovanovich, Inc., 1970), p. 458.)
27. Mumford, "Authoritarian and Democratic Technics."
28. Ellul, *The Technological Society*, p. 224.
29. Lewis Mumford, *My Work and Days: A Personal Chronicle* (New York, NY: Harcourt, Brace and Jovanovich, 1979), p. 528.
30. For an excellent example of such "Luddite shaming," see Robert D. Atkinson, "The 2014 ITIF Luddite Awards," January 5, 2015: www.itif.org/publications/2015/01/05/2014-itif-luddite-awards
31. E. P. Thompson, *The Making of the English Working Class* (New York, NY: Vintage Books, 1966), p. 548.
32. Jennifer Gabrys, *Digital Rubbish: A Natural History of Electronics* (Ann Arbor, MI: The University of Michigan Press, 2013), p. 129.
33. Naomi Oreskes and Erik M. Conway, *The Collapse of Western Civilization: A View from the Future* (New York, NY: Columbia University Press, 2014), pp. 1–2.
34. Ibid., p. 7.
35. Dougald Hine and Paul Kingsnorth, Uncivilization: The Dark Mountain Manifesto (The Dark Mountain Project, 2013), p. 9.
36. Günther Anders, "Theses for the Atomic Age" in Günter Bischof, Jason Dawsey and Bernhard Fetz, *The Life and Work of Günther Anders: Émigré, Iconoclast, Philosopher, Man of Letters* (Innsbruck: StudienVerlag, 2014), pp. 189–90.
37. "Nasa: Earth is warming at a pace 'unprecedented in 1,000 years'," *The Guardian*, August 30, 2016: www.theguardian.com/environment/2016/aug/30/nasa-climate-change-warning-earth-temperature-warming
38. See Sasha Lilley, David McNally, Eddie Yuen, and James Davis, *Catastrophism: The Apocalyptic Politics of Collapse and Rebirth* (Oakland, CA: PM Press, 2012).
39. Weil, *Gravity and Grace*, p. 70.

7

A Critique of Mastery and an Ethics of Attunement: From *Spe Salvi* to *Laudato Si'*

Frank Pasquale

I INTRODUCTION: THE STAKES OF EMBODIMENT

Behind any particular ideological stance there lies a worldview, philosophy, metaphysics, imaginary, or mythos, which either answers or tries to deflect foundational questions about the nature and purpose of human existence.[1] In advanced (and advancing) economies, adherents to now-dominant ideologies of cost–benefit analysis and technocracy have tried to evade or avoid such larger questions in the name of neutrality and pragmatism. Elites assure us that the main contours of the present flow of events are largely unalterable; where change is possible, it should be little more than tinkering to better reflect the positive and negative externalities of market actors. Whatever more robust ethical systems may demand for the far future, they can have little effect in the near term. "Ought" implies "can," and there is no alternative to current forms of economic growth, financialization, increased trade, and technological advance.

This worldview (for now, for convenience's sake, let us call it neoliberalism), can try to domesticate the insights of a document like *Laudato Si'*, by characterizing its deep concerns about environmental degradation and global poverty as best met by technocratic policy and marketized development.[2] William Nordhaus, for instance, complains that "Laudato Si' does not recognize the fact that environmental problems are caused by market distortions rather than by markets per se."[3] He believes that a "market-based solution, particularly carbon pricing, [is] the only practical policy tool we have to bend down the dangerous curves of climate change and the damages they cause," and faults Pope Francis for failing to advocate for such a commonsense solution.[4]

But Nordhaus, like many other readers of the encyclical, misses its deeper point. Catholic Social Thought (CST) does not share the main presumptions of neoliberalism. Rather than beginning an analysis of the social order with markets and the desires of present individuals, it starts with a more open questioning of the origins of those institutions and preferences, and the legitimacy of the practices that inculcated and supported them. *Laudato Si'* in particular is concerned with the *intrinsic value* of nature, not its instrumental use as ecosystem services.[5] A moral philosophy premised on *institutions* and *intrinsic values* will support policy very different than that arising out of the *individualist* and *instrumentalist* worldviews that dominate the self-images of our age.[6]

Laudato Si' challenges a faith in well-governed markets and technological progress as mutually reinforcing cure-alls for environmental devastation and climate change. Without a change in the hearts of those using tools, technological advance is just as prone to accelerating environmental devastation as it is to deflecting it. Crucial steps toward averting disaster will require a confession of the sinfulness of the path that has led us to the present conjuncture, and a recognition of the interdependence and *interbeing* of humans and the planet we inhabit:

> [The earth] now cries out to us because of the harm we have inflicted on her by our irresponsible use and abuse of the goods with which God has endowed her. We have come to see ourselves as her lords and masters, entitled to plunder her at will. The violence present in our hearts, wounded by sin, is also reflected in the symptoms of sickness evident in the soil, in the water, in the air and in all forms of life. This is why the earth herself, burdened and laid waste, is among the most abandoned and maltreated of our poor; she "groans in travail." (LS 2)

Rather than "master[ing]" the earth, we are called to another relation with it – realizing that we "are dust of the earth; our very bodies are made up of her elements, we breathe her air and we receive life and refreshment from her waters." (LS 2) I call this realization (and the sense of moral obligation shaped by it) an ethic of attunement.[7]

Attunement does not command a worship of the earth, or a prioritization of the natural world over urgent human projects. Nor does it offer an obvious or unchanging path to a permanent state of harmony between persons and nature. But an ethical stance of attunement does commend forms of receptivity and sensitivity to the beauty of nature, particularly an appreciation of nature on its own terms, rather than as a standing reserve of resources to be deployed to advance our own purposes. The musical resonance of "tune" in "attunement" is intentional: though "music as an autonomous production is a [relatively] recent invention," we can all sense the absurdity of trying to reduce the "message" of a Bach cantata or Mozart symphony to a sequence of words or equations.[8] Ideally, we listen with an open heart and mind.[9] Indeed, the term "message" itself is hopelessly inexact in presuming the nature of the effect of such art. Its value is incalculable, however much banal calculation, regulation, and bargaining power may interact to shape the price of a piece of music, or the royalties due a musician.[10]

Is nature in this way similar to culture: a source of value and values incommensurable with monetary calculation? Recent CST features a consistent concern with the environment and humankind's place in it. Pope Francis's *Laudato Si'* marks both continuity with extant CST on environmental matters, and an advance growing out of it. The continuity is remarkable in Pope Francis's recurrent critique of aspirations to technological mastery. The advance lies in a profound ethics of attunement that challenges not just the adequacy of present, weak environmental protections, but also the dominant political and economic orders that enframe nature as little more than a pool of resources for human use and enjoyment.

Part II of this chapter explores the continuity between *Laudato Si'* and Pope Benedict's perspectives on environmental topics. Part III explains how Pope Francis's *Laudato Si'* has now complemented extant Catholic critique of value-free mastery of nature with a profound account of *attunement* – that is, the fit between human nature and the natural world we are part of. Part IV explains how this philosophical anthropology has implications for the philosophy of environmentalism, and for policy.

II CONTINUITY: THE CRITIQUE OF MASTERY

Pope Benedict's encyclical *Spe Salvi* develops a critique of "progress-oriented" rationalism, particularly when that kind of instrumentalism amounts to little more than "faith in progress as such."[11] Benedict subjects the Baconian worldview – whereby nature must be subdued to serve human ends – to a sobering reconsideration. "Science can contribute greatly to making the world and mankind more human. Yet it can also destroy mankind and the world unless it is steered by forces that lie outside it," Benedict reminds us, critiquing an aspiration to scientific mastery – whether of nature or persons – as at best incomplete, and at worst disastrous, if it is unleavened by an eschatological perspective on the nature and purpose of human existence. As Pope Francis explains,

> My predecessor Benedict XVI likewise proposed "eliminating the structural causes of the dysfunctions of the world economy and correcting models of growth which have proved incapable of ensuring respect for the environment" . . . Pope Benedict asked us to recognize that the natural environment has been gravely damaged by our irresponsible behaviour. The social environment has also suffered damage. Both are ultimately due to the same evil: the notion that there are no indisputable truths to guide our lives, and hence human freedom is limitless. We have forgotten that "man is not only a freedom which he creates for himself. Man does not create himself. He is spirit and will, but also nature." With paternal concern, Benedict urged us to realize that creation is harmed "where we ourselves have the final word, where everything is simply our property and we use it for ourselves alone. The misuse of creation begins when we no longer recognize any higher instance than ourselves, when we see nothing else but ourselves." (LS 6)

A willingness to view nature as a "standing reserve," a resource for extraction, also helps drive instrumentalist attitudes toward persons as "human resources" for exploitation.[12]

But how do we know what aspects of nature to value – or how much to value them? Respect for nature encompasses a wide spectrum of actions. Jainism commends *ahimsa*, a principle of non-violence extended to all living things.[13] Some strict Jains avoid eating root vegetables, out of respect for the integrity of living things.

The Catholic Church has not seen fit to mandate such guidance for individuals. But it has offered, in CST, a broad set of principles to govern societal relationships with the environment. To fully understand its position, it is helpful to contrast it with alternative approaches to widespread environmental devastation and climate change.

As Anthony Galluzzo has observed, "eco-modernists" "see in the global ecological crisis either a logistical problem to be corrected by some technological deus ex machina – geoengineering or atmospheric reconstitution – or an opportunity for human beings to assume a position of god-like control over the planet."[14] Transhumanists take this position even further, prescribing genetic engineering as a way of producing humans who will be resilient enough to survive future climate shocks.[15] Some science fiction works even endorse unapologetic worship of science itself as a key to the advance of the human condition. For example, in Cixin Liu's novel *The Three Body Problem*, the narrator flatly describes all religions as false, but locates in the scientific process itself – basic research – the key to civilizational advance.[16] In this worldview, ecological activists interfere with the nature and destiny of humankind by conspiring to stop scientific progress.

Laudato Si' takes a much more measured and skeptical approach to technological interventions designed to ameliorate climate change. Too often, in order to solve one problem, we create many others.[17] "We seem to think that we can substitute an irreplaceable and irretrievable beauty with something which we have created ourselves," Francis observes.[18] In its deep respect for biodiversity, the encyclical implicitly critiques aspirations to human mastery of the environment. Some losses are irreversible; they cannot be made up for via simulation, replication, or the blandishments of virtual reality.[19]

In this way, Francis contradicts some of the most important self-images of our age.[20] From a computationalist and instrumentalist perspective, the environment exists as a standing reserve for human exploitation. The parallel development of sciences of human and natural resource management reflect these attitudes. On a reductionist logic, genomic sequences of endangered species should, if stored properly, extinguish legal obligations to preserve them, because the critical information they contain is reflected in their genetic code. Read in light of *Laudato Si'*, this kind of reductionism is wrongheaded in at least three ways.

First, obligations run both ways – not merely from the natural to the human, but from the human to the natural. "It is not enough to think of different species merely as potential 'resources' to be exploited, while overlooking the fact that they have value in themselves." (LS 33) Second, concentration on the value *to humans* of any particular part of the ecosystem can obscure its role in supporting the whole.[21]

Finally, and critically, deep reflection is necessary to discern what Francis calls "the real value of things":

> [W]e need to reject a magical conception of the market, which would suggest that problems can be solved simply by an increase in the profits of companies or individuals. Is it realistic to hope that those who are obsessed with maximizing profits will stop to reflect on the environmental damage which they will leave behind for future generations? Where profits alone count, there can be no thinking about the rhythms of nature, its phases of decay and regeneration, or the complexity of ecosystems which may be gravely upset by human intervention. Moreover, biodiversity is considered at most a deposit of economic resources available for exploitation, with no serious thought for the real value of things, their significance for persons and cultures, or the concerns and needs of the poor. (LS 190)

This passage suggests a critique of the outlook of an economistic mindset that would commend putting all of nature to its "highest and best use." Whether implemented via markets or the state, such a perspective all too often reflects the rule of the strongest. In the wilderness it is particularly suspect.

Environmental policy entails inevitably spiritual dimensions once the very long-term fate of the planet, and people on it, are taken into account. Material and monetary determinations cannot accommodate the range of possible fates in the far future. Consider, for instance, the centrality of discount rates to many standard, neoliberal discussions of the degree of aggressiveness necessary to climate change reduction strategies. As the Stern report noted, the severity of the climate change problem, when presented as the discounted present value of the suffering of future generations due to carbon pollution, can vary wildly depending on the degree to which we discount the value of future generations' experiences. Discounting is the flip side of the so-called "miracle of compound interest": to the extent that we think it natural that a few hundred dollars saved today will compound into a great fortune in a few hundred years, the discounters claim, we need to recognize that even a great fortune in the future could be outweighed by few hundred dollars of cost today. If we believe in accelerating technological advance, we again may not be terribly worried about a warmer world in 2100, assuming that scientists a few generations hence will be capable of geo-engineering or innovations in adaptation.

Duties to future generations do animate the cost–benefit analysis in discount rate discourse, but these duties become ever fuzzier the more one concedes the possibility of very different living conditions in the far future.[22] The discount rate debate is fated to become even more complex and stalled if it focuses entirely on human preferences as policy's *summum bonum*. A more complex ethical outlook, encompassing human duties to preserve nature, is a better framework for policy discussions. And it would likely be a far richer foundation for normative discourse, given the inevitable difficulties in fairly assigning a monetary value to unpredictable scenarios.

At its most solipsistically anthropocentric, faith in technological advance as a "solution" to climate change would commend future creations of virtual reality simulations of any landscape presently existing, to the extent that faithful simulation of its visual, auditory, olfactory, gustatory, and haptic effects is possible. But such an approach, to the extent it expressed respect for nature at all, would merely be an expression of respect for the nature of a given snapshot in time as experienced by humans. *Laudato Si'* instead commends a different type of respect: a willingness to slow, redirect, or change the nature of the footprint of human activity in order to permit the co-evolution of flora and fauna in spaces less influenced by human activity than, say, a city or housing complex.

For some authors, this type of respect can smack of condescension or futility: what good is it to assemble some theme parks of relative stability in a world pervasively shaped by rising carbon dioxide levels? Aren't these spaces of "wilderness" paradoxically anthropogenic, given the inevitably *human* decision to protect them from the (most direct effects of) human habitation? But we should not allow the perfect to be the enemy of the good. While we can no longer aspire to zero impact in any part of the world, we can resolve for a less baleful influence on plants and animals, and their evolution over time.

III ATTUNEMENT AS ROOTEDNESS, RECEPTIVITY, AND RECTIFICATION

In its repeated calls for concerted action to address poverty and environmental disruption, *Laudato Si'* echoes themes of past encyclicals. But it tends to go beyond them in two important ways. First, it takes seriously the neoliberal insistence that economic uplift for the world's least advantaged (and especially the bottom billion or so individuals living on less than $2 a day) may well require further environmental risks. Second, it offers an alternative to such "tragic choices," prescribing certain forms of attunement to nature, and going beyond the negative program of eschewing mastery. These forms of attunement range from the personal (encouraging the relatively wealthy to adopt lifestyles that reflect the lesser environmental impact of the relatively poor), to the political (promoting platforms to tax carbon fairly and invest in renewable energy), to the economic (slowing down cycles of production and consumption that waste resources).

An ethic of attunement also demands recognition of the intertwining of the personal, the political, and the economic. Ethical dispositions of rootedness, receptivity, and rectification help here. Rootedness signifies an attachment to place and purpose, transcending short-term economic demands for worker mobility and labor market flexibility. Receptivity is a willingness to see and treat the natural world as a source of claims upon us – rather than as a standing reserve of resources for our use. A commitment to rectification requires us recognize and scale back some of humanity's greatest offenses against the natural world, ranging from climate change to improper waste disposal to overfishing.

Before elaborating each of these ethical stances, I should address a philosophical problem posed by scientific developments in climate science. A century ago, it was much easier to divide the world into the given and the made, the natural and the artificial, wilderness and developed land. Global climate change now ensures that no part of the earth is untouched by human activities. Bill McKibben called the resulting world "Eaarth," to distinguish it from the prior, more natural one.[23] Jedediah Purdy's book *After Nature: A Politics for the Anthropocene*, observes that "there is no more nature that stands apart from human beings . . . In every respect, the world we inhabit will henceforth be the world we have made.."[24]

Nature is not static, and human influence upon it has already been immense. Thus Celia Deane-Drummond worries that Pope Francis's "interpretation of ecological relations as harmonious relations, which seems to emerge from his particular theological commitment to the value of peace, reflects a specific understanding of ecology in terms of stable relationships that is no longer in vogue among ecological scientists."[25] Earth has not been a pleasant and accommodating home, even for long-standing ecologies. Given the deep history of Ice Ages and other massive climate disruptions pre-dating humanity, the Medea Hypothesis is as plausible as the Gaia Hypothesis.[26]

Despite these worries, Francis's valuation of nature does not shade into romanticization, nor reify any particular version of landscape, seascape, or skyscape as a beau ideal that humanity must strive to restore. Rather, the ethic of care in *Laudato Si'* translates into a sensitivity to time and speed – an affirmation of reflective and considered alteration of habitat. A change in land use or atmosphere may transpire over 1,000 years peacefully; but if it suddenly occurs in a single year, it may be catastrophic. We might call Francis's view of the ideal relationship between persons and nature an ethic of deceleration: a plea to give natural processes an opportunity to adapt to incremental change, rather than imposing upon them sudden and massive human impacts.[27]

This ethic of deceleration may seem to be in tension with other aspects of the encyclical – such as its concern about inclusive development and redistribution to rescue hundreds of millions of persons from extreme poverty. Our primary model of poverty-alleviating development over the past few decades has been China's – a wrenching marketization and industrialization that has caused some of the worst environmental problems in the world, while vastly improving the living standards of hundreds of millions of people. *Laudato Si'* should be read as a call for all people of good will to develop less environmentally damaging versions of economic advance, such as those envisioned in plans for a Green New Deal. Such concern for the "least among us" should be embedded in the ethics of rootedness, receptivity, and rectification discussed below.

ROOTEDNESS

In an era of globalization, cosmopolitanism has eclipsed communitarianism as an ethical aspiration. Yet both will always have a place in CST, as expressions of the aspirational moral character of a church that both makes universal appeals and also realistically counts itself as one of many long-standing faiths. In terms of religious and social institutions, a proper sense of rootedness requires an appreciation of the persons and institutions to whom one owes one's deepest values. In the normal case, a person will owe a lasting and deep partiality and devotion to, say, one's parents, over and above other people. That is not a defect of partiality, but a frank recognition of how much one owes them.[28] Similarly, only in a society bewitched by an extreme individualism would one presume the possibility of stepping aside, at some age of reason, from all one's present ethical and social commitments, to scrutinize all such commitments in light of a universal reason. There will always be some substrate of values behind the ones questioned, and that foundation is at least in part a legacy of locality.

Part of rootedness is a willingness to engage in the here and now, rather than escaping into an increasingly alluring digital world.[29] There are more than elective affinities among decisions to, say, parent by iPad, live off a stock portfolio invested in distant lands, socialize mainly via tweets and Facebook posts.[30] Each of these decisions reflects a distance from the here and now – often thanks to an economic logic dictating the most "productive" use of one's time. Indeed, the growing prominence of remittances in familial survival strategies shows just how brutal the logic of uprooting can be in a globalized capitalist economy. Of course, some economists and attorneys applaud constant "labour mobility" as a cornerstone of efficiency and opportunity – but tend to do so from the safe remove of tenured academic posts, where they can assure that their own families do not have to endure the dislocation and disruption they prescribe for others.

In terms of environmental concern, care for the world should be rooted in love of some particular aspects of the natural world, or place – its flora and fauna, patterns of weather, mix of earth and water. Implied in a love of nature, but not often articulated enough, is a sense of appreciation for these things as such – and not simply as a genetically determined "biophilia," or one particular instance of a kaleidoscope of potential interactions between a particular set of genetic material and a particular environment, but as a conscientiously embraced, fixed point of enduring value. A consumerist mindset, as applied to nature, would leave us like the residents of Italo Calvino's imagined city Leonia, appreciative of our surroundings merely as a salve for boredom.[31]

In *The Need for Roots*, Simone Weil observes that "The notion of obligations comes before that of rights, which is subordinate and relative to the former."[32] Weil glosses the point by explaining that, "A right which goes unrecognized by anybody is not worth very much"; obligations make rights effectual. *The Need for Roots*, as a whole, suggests an additional meaning: that one's sense of obligation ideally informs how one exercises rights. Before asserting oneself as a rights-holder, it is often wise to reflect on one's obligations. These obligations are, at their best, a set of ethical roots that simultaneously restrict and enable, constrain and constitute, one's life plans.[33] The metaphor of roots itself comes from nature, and to the extent that it has meaning for us, a proper sense of gratitude to its origin commends a rootedness in the natural world to complement our attachment to family, party, nation, club, or association.[34]

The value of nurturing communities is a theme of Pope Francis's thought. *Laudato Si'* has an almost heroic sensibility about it when it comes to the power of faith, hope, and love to unify and strengthen communities. For example, Francis's experience in the slums of Latin America shines through in these lines:

> A wholesome social life can light up a seemingly undesirable environment. At times a commendable human ecology is practised by the poor despite numerous hardships. The feeling of asphyxiation brought on by densely populated residential areas is countered if close and warm relationships develop, if communities are created, if the limitations of the environment are compensated for in the interior of each person who feels held within a network of solidarity and belonging. In this way, any place can turn from being a hell on earth into the setting for a dignified life. (LS 149)

One should contrast this stance with Milton's Satan's famous lines on the mind's capacity to "make a Heaven of Hell, a Hell of Heaven."[35] The defiance and arrogance of those lines emerges from a solipsistic ontology. In the poem, Satan underscores mental experience as fundamental. Moreover, it is *his* mind that matters. In Francis's telling, it is a community, a "a network of solidarity and belonging," that can rescue the human spirit from desperation and desolation in the worst of circumstances.

Laudato Si' is, then, a recognition of the fragility of well-being, if not of goodness.[36] Our ability to flourish as human beings – a cornerstone of natural law theory – is inevitably both limited and enabled by our social surroundings. In the limit case, we can imagine children born into conditions of privation – denied, first, proper sustenance and medical care, and then educational, cultural, and spiritual realization, and connections to nature. Each of these deprivations is a form of social sin.[37] The hope in *Laudato Si'* is that a commitment to respect an increasingly vulnerable natural world will also lead to more social respect and regard for vulnerable human beings.

RECEPTIVITY

In the poem *Archaic Torso of Apollo*, Rilke evokes the transformative power of great art: after deep reflection on a Greek sculpture, the viewer is imagined, or called upon, to say to him or herself: "You must change your life." The poem resonates to this day because we can all think of some particularly moving work of art or literature that changed us – illuminating a better way of living, or at least alerting us to certain self-defeating patterns of behavior we had never noticed before. These works embed themselves into our sense of what makes life worth living – of what, on considered reflection, we would like to be remembered for.

But such a process of self-formation is only possible for us when we are somewhat flexible and open. That is one reason why so much ethical and aesthetic formation happens in youth, when our lives' paths and vocational routines are not yet set. Still, sensitive and inquisitive souls, however set they are on some goal, leave open some time for reading and viewing things of intrinsic (and not simply instrumental) importance. They are receptive to aspects of the world, and do not simply seek to instrumentalize it to suit their own pre-existing ends.

The question that a work like *Laudato Si'* raises is: are there natural "texts," experiences of forest, mountain, beach, and desert, inter alia, that should also be paradigmatic episodes of formation for youth – and ongoing engagement for adulthood? How do we remain receptive to the beauty of nature, and to its inherent demands on us? In A *prayer for our earth,* at the end of the encyclical, Francis offers this formulation as an initial step toward awareness:

> All-powerful God, you are present in the whole universe
> and in the smallest of your creatures.
> You embrace with your tenderness all that exists.
>
> ***
>
> Teach us to discover the worth of each thing,
> to be filled with awe and contemplation,
> to recognize that we are profoundly united
> with every creature
> as we journey towards your infinite light.

Receptivity is an aspect of the unity prayed for here – to enjoy solidarity with creation, an exhilaration at its splendor and majesty, pain at its needless squandering at the hand of misguided or excessive development, and more subtle, bittersweet emotions (at, say, the sprout of a morning glory through a cracked sidewalk in a nearly paved-over city). As the growing popularity of centering prayer suggests, spiritual practice creates the space in mind and heart for these experiences, and is reinforced by them.

Of course, the Catholic Church itself can do more to inculcate a spirit of receptivity, of welcoming, as well – and would do well to do so as a step toward modeling this attitude for others, as a seamless garment of respect and concern. Pope Francis has gently guided the Church in this direction by changing its tone on the rights of sexual minorities.[38] Gay, lesbian, and transgender individuals are part of nature (and not deviations from it), however encrusted natural law theory may be with harsher interpretations of such status and conduct.[39] Dramatically evoked by films like *The Twilight of the Goulds*, or poems like Gerard Manley Hopkins's *Pied Beauty*, this ethic of affirmation is of a piece with spiritual practices to inculcate compassion and understanding. As Hopkins put it,

> Glory be to God for dappled things—
> For skies of couple-colour as a brinded cow;
>
> ***
>
> All things counter, original, spare, strange;
> Whatever is fickle, freckled (who knows how?)
> With swift, slow; sweet, sour; adazzle, dim;
> He fathers-forth whose beauty is past change:
> Praise Him.[40]

This kind of pluralism and affirmation of the mystery and diversity of reality *as it exists* – not as some idealized or standardized vision – is a hallmark of a spirit that goes beyond tolerance and respect, toward a kind of grace and understanding that is so necessary in our age.

RECTIFICATION

In the American Museum of Natural History, a 2010 exhibit explored the future of climate change. The exhibit chronicled rising sea levels, freak storms, and endangered aspects of the natural world, increasingly threatened by ever-warmer weather. But it ended on a high note. Posters titled "Adaptation to Climate Change" described how human beings could change their ways of life – and even themselves – to adjust to a dramatically hotter planet.[41] For example, pioneers could start colonizing caves and develop underground spaces that were cooler than the surface of the planet. Humans might need to be shorter to handle the new environs – but genetic engineering can handle that.[42] Lack of sunlight may be an issue, too – but perhaps a desire for solar exposure, like height, could be modulated in future humans, too.[43] Technicians may eventually "geoengineer" climate itself, planning an optimal mix of sun and clouds for every spot on earth. That would potentially set in motion a spiraling set of interventions whose ultimate impact, decades later, is certainly no more knowable than the weather three weeks from now.

Such "adaptations" are jarring to contemplate. But are they morally or legally problematic? A few routes of critique suggest themselves immediately. Disability rights activists have condemned social mores that require individuals to adapt themselves to a given, human-made social order – rather than adjusting that order to accommodate human beings. The United States' Endangered Species Act exemplifies another way of valuing what *is*, rather than insisting on its adaptation. It stops projects that would eliminate or seriously degrade the habitat of a species that is in danger of extinction.[44] Such decisions may endanger certain kinds of jobs or development. But the larger ethical challenge is to avoid the necessity of that kind of trade-off.[45] It is to *rectify* damage already done, rather than to change the game entirely.[46]

Certainly there are changes that can be made to *socio-economic systems*, so that they do not conduce to the creation of atmospheric carbon that is now threatening to change the environment so much that we must engineer ourselves toward post-humanity.[47] It is not human life itself, but rather, certain mutable ways of organizing our economic and political life, that lead to climate change. Some terrible effects of climate change can be averted if global leaders manage to develop and deploy alternatives to fossil fuels.

Will they do so? Aggressive lobbying and disinformation campaigns from the fossil fuel lobby (among other purveyors of business-as-usual) have stalled reform for decades.[48] Beyond *realpolitik* and regulatory capture, there are conceptual blind spots in contemporary policy analysis that diminish the urgency of reacting to climate change. The dominant mode of policy analysis in the twenty-first century – cost–benefit calculations – cannot address the types of problems raised by climate risk decades from now, once the possibility of adapting future humanity to a world changed utterly by current humans is raised. Cost–benefit analysis presumes a stable, typical subject to enjoy benefits or suffer costs. Once the subject is rendered mutable, costs and benefits themselves change.

Technocratic policy analysts try to discipline this and all manner of other uncertainty into the question of "discount rates." Discounting the future has become common because, given certain assumptions about inflation and the time value of money, a dollar today is worth more than a dollar twenty years from now.[49] A high discount rate implies a much lower value for the welfare of future humans than a low discount rate. Adaptationist optimism excuses high discount rates: however wrecked we may leave habitats and ecosystems, our ancestors' technological prowess will somehow put things to right. Or, even more insidiously: that future humans or cyborgs will be unperturbed the destruction of habitat, given the right mix of virtual reality and psychopharmaceuticals.[50]

Policy science loses all mooring once such possibilities arise. Thus, in practice, *Laudato Si'* confirms Douglas Kysar's profound critique of cost–benefit analysis as an enabler of "regulation from nowhere."[51] However broad the utilitarian calculus of costs and benefits, pain and pleasure may be, we must recognize that such primitive evaluative tools arise out of a primordial experience of embodiment in nature. Preserving some critical mass of that nature (on a macro level, and at the level of the human person), is a precondition for making sense of claims about costs and benefits. As Pope Francis wisely suggests, the ultimate freedom lies in recognition of our deepest and most enduring duties – including the obligation to pass to future generations a natural world recognizably continuous with that which shaped the past which gives our experience meaning.[52] Rectifying our present, broken relationship with nature amounts to a better framing narrative for climate change response, than any effort to create a new nature.

In his book *Finance and the Good Society*, Nobel laureate Robert Shiller looked at what he considered to be the bright side of the BP oil spill: "to the extent that all risks were insured properly ... then there was no real and lasting suffering," he stated.[53] Compromised ecosystems, disrupted lives, countless hours of cleanup – all smoothed away by the mellifluous flow of properly allocated money. This tone-deafness to the suffering or dislocation of wildlife in the gulf is a startling reminder that even progressive economists can devalue the natural world.

Earth still has many beautiful and undespoiled spots, but also many devastated ones. The silence of Chernobyl stands as a lasting reminder of humankind's arrogance. It is a sacrifice zone, in many ways unremediable, and a lasting reminder of the high stakes of environmental risk taking. As earthquakes rock Oklahoma thanks to fracking activity, other sacrifice zones may emerge – even as many energy options with far less damaging impacts are neglected by governments captured by fossil fuel lobbies.[54] Even if certain environments can never be restored, we can work to rectify the misdeeds that doomed them – for example, by advancing societal adoption of less destructive energy options.

But we should also acknowledge the tension between rootedness and rectification, when it takes an overly technologized form. For example, in some areas suffering from bee colony collapse disorder, roboticists have proposed the deployment of tiny drones to act as substitute pollinators. Such drones may be great pollinators – but do their programmers also intend them to mimic all the other roles bees play in ecosystems? Do we even know the full extent of such roles? There is a wisdom to precautionary principles for ecological interventions.

Nor should aesthetic and spiritual attachments to nature be dismissed as soft-headed romanticism. On a personal level, I would experience a garden buzzing with RoboBees as a tragedy – a site of loss. Admittedly, a child born in a bee-less world might never know what he had missed. But that potential ignorance of the past – or indifference to parts of it now fading – is not a reason to consign our present to radical change without exploring all possible avenues for respecting the earth, by decelerating destructive development now trashing our lands, waters, and skies. To be sure, there is no eternal and pristine landscape to save – the environment evolves over time. But that span of time was immense compared with the pace of technological advance today. As *Laudato Si'* reminds us, nature is not simply a standing reserve to be manipulated to our ends. Its intrinsic value is distinct from its instrumental use as ecosystem services or temporary camp site.

Accelerationists, who welcome rapid technological change in both social relations and the natural world, may find such a point of view quaint or mystical. But it is hard to see how they could consistently cordon off humanity itself from the type of rapid manipulation they would prescribe for the planet on which we evolved. Why not solve the problem of scarcity by genetically engineering human beings to be satisfied with the status quo? Why not promote deeply egalitarian social relations by medicating away envy or strivings for superiority? To guarantee a truly humane future, we will need institutions and economies that can keep pressures toward the mechanization, standardization, and diminution of human experience in check. Respect for the stability of what is – be it nature or human nature – will be indispensable.

IV CONCLUSION: BEYOND MECHANISM

From a purely mechanistic, "objective" perspective, there is a seamless transition from physics and biology to more human sciences. Joseph Vining evokes this transition in his *From Newton's Sleep:*

> Survival and increase, world without end. Quite empty these are. But familiar. These are the objects of man's activity in the eyes of the classical economist or political scientist, of life itself in the eyes of the classical biologist. These comprise indeed the very definition of life in science.
>
> Being empty, they can be translated from objectives into consequences and thus link the study of life and man with the study of physical science. It is clever, the way the economic and biological visions fit together. Which came first? Seeing man as seeking only survival and increase, more and more aggrandizement, preferably of an abstraction, more dollars, genes, power, swallowing up, spreading over the whole world if not stopped from without? Or seeing animate nature as without purpose?[55]

Technocratic cost–benefit analysis may be a useful guide in certain narrow, short-term contexts. It can warn us not to build a new home in a flood plain, for example, and to reallocate those resources to some higher and better use. But as the scope of time, space, and expense opens up, the method loses its mooring. The longer the term of our vision, the more basic ideals and values must come to the fore. If we want a future somewhat recognizable to the world of today, we must buttress the institutions capable of inculcating certain habits of mind and character that respect intrinsically valuable aspects of our experience.

Laudato Si' powerfully demonstrates the Catholic Church's commitment to preserving respect and concern for the environment for posterity. And it also shows how better-known Catholic emphases on the preciousness of human life imply a lasting value for the world that has enabled and supported that life, both on the time scale of evolution, and in the more short-term immediacy of preserving habitat in environments most threatened by climate change. Pope Benedict XVI spoke of an "ecology of man," based on the fact that "man too has a nature that he must respect and that he cannot manipulate at will" (LS 120). This core idea of the necessary critical mass of stability in *nature*, and *human nature*, is one that should animate any long-term cost–benefit analysis, lest it degenerate into the type of subjectivism discredited long ago by, *inter alia*, Nozick's thought experiment on the experience machine.[56]

Given the powerful critiques of economism and scientism in works like *Laudato Si'*, academics should give up on developing a science of policy cleansed of all biases and contestable values – a *View from Nowhere*, in Nagel's evocative phrasing.[57] Explicitly acknowledging ultimate questions in environmental law (and the ultimate answers to them encoded in any reasonably comprehensive policy framework) may seem like a step down a slippery slope of sectarianism. But the alternative is an even more troubling drift away from the foundations of recognizably human values.

We must also reconsider and expand the goals of technology policy. Innovation is important, but maintenance of the natural and man-made infrastructure we already enjoy is foundational.[58] It makes little sense to defund sewage systems in order to reduce taxes on inventors on the vague promise that they may, one day, develop some heretofore unimaginable self-contained system of water provision and waste destruction. Similarly, before investing in a mission to colonize space, it behooves us to do as much as we can to assure the lasting integrity of the world we evolved in.[59]

Alasdair MacIntyre has argued that "an educated Catholic laity needs to understand a good deal more about Catholic philosophical thought than it now does."[60] *Laudato Si'* helps close that gap by explaining in accessible terms the high stakes of continuing to neglect the preconditions for a stable climate. The encyclical should also reach beyond the "urbi" to the "orbe," or the world as a whole: as a social theory, and a framework of value that transcends the tyranny of present economic patterns. However tempting schemes of geoengineering or genetic engineering may become, we should be very cautious about schemes to "master" human (and non-human) nature. If they are to proceed at all, they should be preceded by a spiritual project of attunement to creation and recognition of what it is that makes us human. *Laudato Si'* is a critical source to inform such a philosophical anthropology.

Notes

1. Richard H. Brown helpfully catalogs such overarching terms of moral and ontological orientation: Richard Harvey Brown, *A Poetic for Sociology: Toward a Logic of Discovery for the Human Sciences* (New York, NY: Cambridge University Press, 1977), pp. 125–26.
2. For succinct definitions of neoliberalism, see Philip Mirowski, *Never Let a Serious Crisis Go to Waste: How Neoliberalism Survived the Financial Meltdown* (New York, NY: Verso, 2013); David Singh Grewal and Jedediah Purdy, "Introduction: Law and Neoliberalism," *Law and Contemporary Problems*, 77 (2014), 1–24.
3. William D. Nordhaus, "The Pope and the Market," *New York Review of Books*, October 8, 2015: www.nybooks.com/articles/2015/10/08/pope-and-market/ Nordhaus recently won the Nobel Prize for his climate work, but may be better remembered for unduly minimizing the risk of climate change at a critical point in human history. Jason Hickel, "The Nobel Prize for Climate Catastrophe," *Foreign Policy*, December 6, 2018: https://foreignpolicy.com/2018/12/06/the-nobel-prize-for-climate-catastrophe/ ("[H]ow do economists get away with believing that these extreme temperatures are somehow okay? Because the Nordhaus model tells us that even the worst catastrophes will not really hurt the global economy all that much. Maybe a percentage point or two at the most, by the end of the

century—much less than the cost of immediate action. . . . [I]f climate breakdown ends up starving and displacing a few hundred million impoverished Africans and Asians, that will register as only a tiny blip in GDP. After all, poor people don't add much 'value' to the global economy [according to this worldview]. The same goes for things like insects and birds and wildlife, so it doesn't matter if global warming continues to accelerate mass extinction. From the perspective of capital, what most of us see as tremendous ethical and even existential problems literally don't count.").

4. Ibid.
5. For an account of environmental economists' efforts instrumentally to value nature, see Frank Pasquale, "Toward an Ecology of Intellectual Property: Lessons from Environmental Economics for Valuing Copyright's Commons," *Yale Journal of Law and Technology*, 8 (2005), 78–135.
6. Alasdair MacIntyre, *Against the Self-Images of the Age: Essays on Ideology and Philosophy* (Notre Dame, IN: University of Notre Dame Press, 1978); Frank Pasquale and Arthur J. Cockfield, "Beyond Instrumentalism: A Substantivist Perspective on Law, Technology, and the Digital Persona," *Michigan State Law Review* 18 (2018), 821–70.
7. I loosely borrow the metaphor from William E. Connolly, *Identity/Difference: Democratic Negotiations of Political Paradox* (Ithaca, NY: Cornell University Press, 1991), 29 ("A third doctrine opposes these two quests for mastery. It sees more than indifference in nature; it discerns a bent or direction in the world to which the self and the community must strive to become attuned. The self becomes more free by becoming more attuned to the deepest purposes inscribed in its community, and the community becomes more free by becoming more attuned to the bent of nature in the self and the world. Freedom involves above all attunement to a higher direction or harmony; it is fulfillment and harmonization.").
8. Jacques Attali, *Noise: The Political Economy of Music* (Minneapolis, MN: University of Minnesota Press, 1985), pp. 1, 3 ("we must learn to judge a society more by its sounds, by its art, and by its festivals, than by its statistics. By listening to noise, we can better understand where the folly of men and their calculations is leading us, and what hopes it is still possible to have."); Charles Taylor, *Sources of the Self: The Making of the Modern Identity* (Cambridge, MA: Harvard University Press, 1989) (describing the power of great art to shape the self that assesses it, and the concept of a hermeneutic circle).
9. When Max Weber said that he was religiously "unmusical," he also evoked some of my sense here – in that musicality denotes a certain compatibility and affinity (that goes beyond or beside skill), to connote one's way of being in the world. We are all more or less "musical" in various spheres of life to the extent they come naturally to us. This root metaphor of "naturally" itself derives from nature, from the stable surround that is our ground of being.
10. It can be hard to make the case directly for such intrinsic worth – until one runs into paradoxes of cost–benefit analysis (including Richard Posner's bizarre back-of-the-envelope calculation of the cost of human extinction at $600 trillion). Richard A. Posner, *Catastrophe: Risk and Response* (New York, NY: Oxford University Press, 2004) (quoted in Clifford Geertz, "Very Bad News," *New York Review of Books*, March 24, 2005: www.nybooks.com/articles/2005/03/24/very-bad-news/). One wonders *to whom* this would be a cost, given that there would be no one left to pay it. Such are the antinomies of methodologically imperialistic cost–benefit analysis.
11. Pope Benedict XVI, *Spe Salvi* (Saved by Hope), November 30, 2007: http://w2.vatican.va/content/benedict-xvi/en/encyclicals/documents/hf_ben-xvi_enc_20071130_spe-salvi.pdf
12. The human results of such instrumentalization are brilliantly invoked in Laurent Cantet's film *Ressources Humaines* (1999; Brooklyn, NY: KimStim, Inc., 2004), DVD.
13. Ara Paul Barsam, *Reverence for Life: Albert Schweitzer's Great Contribution to Ethical Thought* (New York, NY: Oxford University Press, 2008).
14. Anthony Galluzzo, "Neither Promethean nor Primitivist: A Response to Andreas Malm," *Verso Blog*, June 30, 2016: www.versobooks.com/blogs/2743-neither-promethean-nor-primitivist-a-response-to-andreas-malm
15. See also Michael Shellenberger and Ted Nordhaus (eds.), *Love Your Monsters: Postenvironmentalism and the Anthropocene* (United States: Breakthrough Institute, 2011).
16. Cixin Liu, *The Three Body Problem*, trans. Ken Liu (New York, NY: Tor Books, 2014), pp. 393–94. Writing for himself in an author's postscript, Cixin Liu states: "I've always felt that the greatest and

most beautiful stories in the history of humanity were not some by wandering bards or written by playwrights and novelists, but told by science. The stories of science are far more magnificent, grand, involved, profound, thrilling, strange, terrifying, mysterious, and even emotional, compared to the stories told by literature. Only, those wonderful stories are locked in cold equations that most do not know how to read. The creation myths of the various peoples and religions of the world pales when compared to the glory of the Big Bang."

17. LS 34 ("Human beings must intervene when a geosystem reaches a critical state. But nowadays, such intervention in nature has become more and more frequent. As a consequence, serious problems arise, leading to further interventions; human activity becomes ubiquitous, with all the risks which this entails. Often a vicious circle results, as human intervention to resolve a problem further aggravates the situation. For example, many birds and insects which disappear due to synthetic agrotoxins are helpful for agriculture: their disappearance will have to be compensated for by yet other techniques which may well prove harmful. We must be grateful for the praiseworthy efforts being made by scientists and engineers dedicated to finding solutions to man-made problems. But a sober look at our world shows that the degree of human intervention, often in the service of business interests and consumerism, is actually making our earth less rich and beautiful, ever more limited and grey, even as technological advances and consumer goods continue to abound limitlessly.")
18. Ibid.
19. Compare John Horgan, *The Undiscovered Mind: How the Mind Defies Replication, Medication, and Explanation* (New York, NY: Free Press, 1999).
20. MacIntyre, *Against the Self-Images of the Age*.
21. LS 34 ("It may well disturb us to learn of the extinction of mammals or birds, since they are more visible. But the good functioning of ecosystems also requires fungi, algae, worms, insects, reptiles and an innumerable variety of microorganisms. Some less numerous species, although generally unseen, nonetheless play a critical role in maintaining the equilibrium of a particular place.").
22. Brett M. Frischmann, "Some Thoughts on Shortsightedness and Intergenerational Equity," *Loyola University Chicago Law Journal*, 36 (2005), 457–68.
23. Bill McKibben, *Eaarth: Making a Life on a Tough New Planet* (New York, NY: Times Books, 2010).
24. Jedediah Purdy, *After Nature: A Politics for the Anthropocene* (Cambridge, MA: Harvard University Press, 2015).
25. Celia Deane-Drummond, "Laudato Si' and the Natural Sciences: An Assessment of Possibilities and Limits," *Theological Studies*, 77 (2016), 392–415.
26. Peter D. Ward, *The Medea Hypothesis: Is Life on Earth Ultimately Self-Destructive?* (Princeton, NJ: Princeton University Press, 2009).
27. On the problem of the "social acceleration of time," see Hartmut Rosa and William E. Scheuerman (eds.), *High Speed Society: Social Acceleration, Power, and Modernity* (University Park, PA: Pennsylvania State University Press, 2009); Hartmut Rosa, *Social Acceleration: A New Theory of Modernity* (New York, NY: Columbia University Press, 2013). Rosa himself does not endorse deceleration, but it is an orientation to change could inspire an ethical disposition toward nature that reflects the ideals articulated in *Laudato Si'*.
28. Peter Berkowitz, "Other People's Mothers," *The New Republic*, January 10, 2000, 27–37 (on Peter Singer).
29. Cf. Paul Virilio, *Open Sky* (New York, NY: Verso, 1997).
30. Cf. Paul Elie, *The Life You Save May Be Your Own* (New York, NY: Farrar, Straus and Giroux, 2003) (on Walker Percy).
31. Italo Calvino, *Invisible Cities*, trans. William Weaver (New York, NY: Harcourt Brace, 1972). Bauman reflects on the story as a metaphor for "perpetual, compulsive, obsessive, and addictive modernization." (Zygmunt Bauman, Wasted Lives: Modernity and its Outcasts (Malden, MA: Polity, 2004), p. 6). See also Frank Pasquale, "Access to Medicine in an Era of Fractal Inequality," *Annals of Health Law*, 19 (2010), 269–310.
32. Simone Weil, *The Need for Roots: Prelude to a Declaration of Duties Toward Mankind*, trans. Arthur Wills (New York, NY: Putnam, 1952).
33. Cf. John Kekes, "The Meaning of Life" in E. D. Klemke and Steven Cahn (eds.), *The Meaning of Life: A Reader*, 3rd edn. (New York, NY: Oxford University Press, 2008).

34. George Lakoff, *Metaphors We Live By* (Chicago, IL: University of Chicago Press, 1980).
35. John Milton, *Paradise Lost*, Book 1, lines 250–255:

    Hail horrours, hail
    Infernal world, and thou profoundest Hell
    Receive thy new Possessor: One who brings
    A mind not to be chang'd by Place or Time.
    The mind is its own place, and in it self
    Can make a Heav'n of Hell, a Hell of Heav'n.

36. Martha C. Nussbaum, *The Fragility of Goodness: Luck and Ethics in Greek Tragedy and Philosophy* (New York, NY: Cambridge University Press, 1986).
37. *In the Company of the Poor: Conversations with Dr. Paul Farmer and Gustavo Gutiérrez*, ed. Michael Griffin and Jennie Weiss Block (Maryknoll, NY: Orbis Books, 2013).
38. James Martin, *Building a Bridge: How the Catholic Church and the LGBT Community Can Enter into a Relationship of Respect, Compassion, and Sensitivity* (New York, NY: HarperCollins, 2017).
39. Even contemporary natural law accounts all too often fail to acknowledge the persistence and prevalence of same-sex attraction: Stephen Macedo, "Against the Old Sexual Morality of the New Natural Law Theory" in Robert George (ed.), *Natural Law, Liberalism, and Morality: Contemporary Essays* (New York, NY: Oxford University Press, 2001).
40. Gerard Manley Hopkins, "Pied Beauty," in *Poems of Gerard Manley Hopkins* (Robert Bridges ed., 1918).
41. Jane Mayer, "Covert Operations," *The New Yorker*, August 30, 2010 (which includes this vignette).
42. Matthew Liao, "Small People Save the Environment," *Quest Magazine*, June 2012.
43. Jonathan Crary, *24/7: Late Capitalism and the Ends of Sleep* (New York, NY: Verso Books, 2013), on modafinil and other modifications of human needs.
44. Endangered Species Act, 16 U.S.C. § 1531 *et seq.* (2012).
45. Naomi Oreskes and Erik M. Conway, *The Collapse of Western Civilization: A View from the Future* (New York, NY: Columbia University Press, 2014).
46. For the metaphor of "changing the game," I am indebted to Lawrence Joseph's book, *The Game Changed: Essays and Other Prose* (Ann Arbor, MI: University of Michigan Press, 2011).
47. John Paul II, Message to the Pontifical Academy of Sciences: On Evolution, October 22, 1996: www.ewtn.com/library/PAPALDOC/JP961022.HTM
48. Naomi Oreskes and Eric M. Conway, *Merchants of Doubt* (London: Bloomsbury, 2010).
49. Frischmann, "Some Thoughts on Shortsightedness." Of course, our recent experience of negative interest rates on billions of dollars of sovereign debt raises deep questions about discounting methodology. So, too, does the manipulability of interest rates analyzed in modern monetary theory, among other heterodox schools of macroeconomics.
50. Frank Pasquale, "Two Concepts of Immortality: Reframing Public Debate on Stem-Cell Research," *Yale Journal of Law and the Humanities*, 14 (2013), 73–122.
51. Douglas A. Kysar, *Regulating from Nowhere: Environmental Law and the Search for Objectivity* (New Haven, CT: Yale University Press, 2010).
52. Pasquale, "Two Concepts of Immortality," 94, discussing Wallace Stevens's "A Postcard from the Volcano" in *The Collected Poems of Wallace Stevens* (New York, NY: Alfred A. Knopf, 1965), p. 71.
53. To his credit, Shiller did deplore the loss of human lives in the explosion itself – but failed to account properly for the damage to ecosystems and species within them.
54. For the concept of sacrifice zone, see Chris Hedges and Joe Sacco, *Days of Destruction, Days of Revolt* (New York, NY: Nation Books, 2012).
55. Joseph Vining, *From Newton's Sleep* (Princeton, NJ: Princeton University Press, 1994), p. 294. See also Frank Pasquale, "Technology, Competition, and Values," *Minnesota Journal of Law, Science, and Technology*, 8 (2007), 607–22.
56. Robert Nozick, "The Experience Machine" in *Anarchy, State, and Utopia* (New York, NY: Basic Books, 1968), p. 42; Frank Pasquale, "Libertarians Against Subjectivism," *Concurring Opinions*, April 30, 2007: https://concurringopinions.com/archives/2007/04/libertarians_ag.html. As a matter of theory, *Laudato Si'* challenges the rise of anthropocentric relativism (the increasingly common

view that, given humanity's extant influence on climate, there is no baseline of "wilderness" to save, and humankind must be the measure of all things – including geoengineering initiatives). *Laudato* also provides a counterweight to singularitarian interpretations of the theology of Teilhard de Chardin, given their gnostic bias toward imagining the transferability of mind and soul to inorganic matter. For insightful perspectives on the religious aspects of transhumanist thought, see Meghan O'Gieblyn, "God in the Machine: My Strange Journey into Transhumanism," *The Guardian*, April 18, 2017; Charles T. Rubin, *Eclipse of Man: Human Extinction and the Meaning of Progress* (New York, NY: Encounter Books, 2014).

57. Thomas Nagel, *The View from Nowhere* (New York, NY: Oxford University Press, 1986).
58. Andrew Russell and Lee Vinsel, "Hail the Maintainers," *Aeon*, April 7, 2016: https://aeon.co/essays/innovation-is-overvalued-maintenance-often-matters-more
59. Sue Halpern, "The Man for Mars," *The New York Review of Books*, August 13, 2015: www.nybooks.com/articles/2015/08/13/elon-musk-man-mars/
60. Alasdair C. Macintyre, *God, Philosophy, Universities: A Selective History of the Catholic Philosophical Tradition* (Lanham, MD: Rowman and Littlefield, 2009).

PART III

Catholic Social Thought at Work and Play

8

"Truly, Much Can Be Done!": Cooperative Economics from the Book of Acts to Pope Francis

Nathan Schneider[*]

Among the controversies that accompanied the release of Pope Francis's *Laudato Si'* was a perplexing diversity of perceptions regarding the encyclical's mood. This wasn't so much a topic of outright dispute as it was a diversity of assumptions beneath the various salvos of analysis.[1] Some readers experienced the text as being in keeping with Francis's spirit of "the joy of the Gospel" articulated in his 2013 document *Evangelii Gaudium*. "*Laudato Si'* indeed!" proclaimed Michael Sean Winters in the *National Catholic Reporter*. "Our Holy Father has blessed the Church with a document that is accessible to virtually anyone, rich in the collected wisdom of the Catholic faith."[2] Among the staff and fellow travelers of *First Things*, meanwhile, the perspective was quite different. Editor R. R. Reno's headline referred to its "weakness"; Deputy Editor Matthew Schmitz opined in *The Washington Post* on the pope's "dark tone"; *New York Times* columnist Ross Douthat characterized the text as "sprawling," "apocalyptic," and "foreboding."[3]

This latter interpretation was, for me, baffling. The encyclical's title translates as "Praise Be to You" – a quotation from a poem that revels in God's creation. Throughout, alongside the warnings and soundings of alarm, there are signs of progress, of possibility. Francis then concludes with a prayer that is nothing if not an exhortation to hope:

> O Lord, seize us with your power and light,
> help us to protect all life,
> to prepare for a better future,
> for the coming of your Kingdom
> of justice, peace, love and beauty.
>
> (LS 294)

After puzzling over this difference of perception for a little while, I came to a theory that might, at least in part, account for it. (I would prefer to think that the matter can't merely be chalked up to self-described conservatives trying to undermine the message of a pope they perceive as liberal.) Perhaps, I surmised, the signs of progress and possibility that Francis cites aren't universally hopeful because they're not universally familiar. Perhaps some of us aren't clear on what he is talking about, on the reasons for the hope within him, on the long tradition of Catholic thought and practice to which he seems so joyfully to refer.

Consider, for instance, this passage, found in the last third of the encyclical:

> There are no uniform recipes, because each country or region has its own problems and limitations. It is also true that political realism may call for transitional measures and technologies, so long as these are accompanied by the gradual framing and acceptance of binding commitments. At the same time, on the national and local levels, much still needs to be done, such as promoting ways of conserving energy. These would include favouring forms of industrial production with maximum energy efficiency and diminished use of raw materials, removing from the market products which are less energy efficient or more polluting, improving transport systems, and encouraging the construction and repair of buildings aimed at reducing their energy consumption and levels of pollution. Political activity on the local level could also be directed to modifying consumption, developing an economy of waste disposal and recycling, protecting certain species and planning a diversified agriculture and the rotation of crops. Agriculture in poorer regions can be improved through investment in rural infrastructures, a better organization of local or national markets, systems of irrigation, and the development of techniques of sustainable agriculture. New forms of cooperation and community organization can be encouraged in order to defend the interests of small producers and preserve local ecosystems from destruction. Truly, much can be done![4]

It would require a bold feat of interpretation to understand "Truly, much can be done!" as apocalyptic or foreboding. But what is the cause of the pope's enthusiasm? Possibilities. He revels in a multifarious toolkit for the adventure of social and ecological reform. And the direct antecedent to his final exclamation, "new forms of cooperation and community organization," seems to summarize the long litany and punctuate it. As popes often do, when speaking of something new he is also speaking of something old; its simultaneous oldness and newness warrants the confidence of Francis's hope.

This chapter will consider what the encyclical means by "cooperation" as a form of "organization" in depth. This is not merely a vague invitation to collaborate and play nice. Additional references to cooperative organizing in the text, together with related statements elsewhere, make clear that Francis is referring to a specific kind of strategy that he considers to have special promise for social and ecological justice. Cooperation is not the only strategy he recommends, as this lengthy litany makes clear. Yet it is an important one, and has been almost universally ignored in the scholarly and journalistic commentary on *Laudato Si'*.

Economic cooperation[5] is a way of doing business based in democratic ownership and governance, a striving for genuine stakeholdership and stewardship. Cooperatives put the people and communities most affected by an enterprise in control, empowering them to meet their own needs in tandem with the common good. Modern co-ops, unlike investor-owned corporations, generally commit to one-member-one-vote decision-making, rather than allocating more power to those who own more shares, and they aspire to cooperation among enterprises rather than competition. Such enterprises can take various legal forms, but they share an ownership model designed to maximize community benefit, not investor profits.[6] Farmers' cooperatives have proven a dominant model for small-scale agricultural producers the world over, and credit unions consistently provide capital access in communities that investor-owned banks fail to serve. A 2014 United Nations report counted 2.6 million cooperatives, with over 1 billion memberships and more than 12.6 million employees internationally.[7]

This tradition of thought and practice has a long history and deep roots in Christian faith and Catholic tradition in particular. It's a legacy with which Francis has exhibited a robust familiarity. By and large, however, US Catholics seem to know little about cooperative economics or its relationship with their faith, even while church organizations continue to be among its chief supporters.

Cooperativism is, on the one hand, a necessary key for interpreting *Laudato Si'*, which explicitly recommends cooperative enterprise three times and repeatedly uses *cooperation* to describe the proper human posture toward God and creation; it has also been a persistent, under-recognized theme in Francis's worldview and that of Catholic social thought for more than a century. For Francis, this kind of enterprise grows out from the joy of the Gospel and proves the reasonableness of his call for "integral ecology." In what follows, I set out to reintroduce cooperative economics through the eye of Catholic tradition and Francis's own utterances.

FRANCIS (AND BERGOGLIO) ON COOPERATIVE ECONOMICS

Jorge Bergoglio was already a cardinal during Argentina's economic collapse in 2001. Among other improvizations during the crisis, workers in shuttered factories occupied them and kept them running under worker control; during that time, Bergoglio defended the popular uprisings against police repression.[8] He later acted as an advocate for an anti-trafficking organization called La Alameda, which emerged from a textile workers' cooperative and supported people escaping sweatshops and brothels, in part, by helping them to form cooperative businesses.[9]

I spoke by phone with Néstor Escudero,[10] a member of La Alameda, who described how Bergoglio would lend his visibility to protect the organization and those it served from retribution. This was, Escudero believes, an act of mercy on behalf of the vulnerable more than a means of favoring one business model over another. But the cardinal appreciated the business model as well. "We know that he supports co-ops because he has expressed it in meetings with other organizations when we had social movement gatherings," Escudero told me. Escudero also noted that there are reasons to be skeptical; co-op law in Argentina can sometimes become a means of skirting labor law and treating workers unfairly.

Bergoglio's positive perception of cooperative enterprise, however, was already well-rooted. In 2013, as Pope Francis, he responded with satisfaction to reports of co-ops that had managed to maintain their level of employment even during the recent financial crisis. This inclined him to tell a story he dated to 1954, when he was still a teenager. "I heard my father" – an accountant – "speak on Christian cooperativism and from that moment I developed an enthusiasm for it, I saw that it was the way," Francis recalled. He went on:

> It is precisely the road to equality, not to homogenity [sic], but to equality in difference. Even economically it goes slowly. I remember that reflection my father gave: it goes forward slowly, but it is sure.[11]

Two years later, Francis echoed this sentiment at the World Meeting of Popular Movements in Bolivia:

> I have seen at first hand a variety of experiences where workers united in cooperatives and other forms of community organization were able to create work where there were only crumbs of an idolatrous economy. Recuperated businesses, local fairs and cooperatives of paper collectors are examples of that popular economy which is born of exclusion and which, slowly, patiently and resolutely adopts solidary [sic] forms which dignify it.[12]

The question of temporality, actually, has been a persistent preoccupation for Francis. In *Evangelii Gaudium* he articulates four principles that have guided his ecclesiastical life over the years. The first of these – "time is greater than space" – has been a motto for him since at least 1974.[13] This might sound like an esoteric interpretation of physics, but the pope's explanation in *Evangelii Gaudium* is squarely about politics. "Giving priority to time," he writes, "means being concerned about initiating processes rather than possessing spaces."

Although a leader might be tempted to assert her power in the short term to score a victory, this motto insists that fostering healthy process matters more in the long run. It's a principle that the pope put on display during the Synod on the Family proceedings in 2014 and 2015, before which he encouraged bishops to carry out consultations with the faithful. He then called the bishops to Rome for unusually transparent deliberations, to which his final conclusions deferred. Perhaps it would have been more expedient for him simply to assert his position and demand allegiance, but, seeking to inculcate habits of cooperation, he chose to institute a patient and participatory process.

Francis has also demonstrated awareness of the pitfalls that cooperative enterprises can face. In his 2015 address to the Confederation of Italian Cooperatives, he offered a warning:

> When money becomes an idol, it commands the choices of man. And then it destroys man and condemns him. It renders him a servant. Money at the service of life can be managed in a just way by a cooperative, if however, it is an authentic, true cooperative, *where capital is not in command over men but men over capital*.[14]

However abstract, this speaks to a very specific phenomenon among large, long-standing co-ops that over time shed their cooperative identity by operating like conventional capitalist firms. He denies such firms the easy satisfaction of claiming the cooperative label. In an address to the same body the following year, Francis further develops this distinction between the "true cooperative" and others: "Counter the false cooperatives," he said, "because cooperatives must promote an economy of honesty."[15] Here, the honesty he's talking about involves solidarity within ("Manage the cooperatives truly cooperatively – that is, involving all") and service to the common good ("Participate actively in globalization in order to integrate – in the world – development, justice, and peace"). These concerns highlight a persistent tension between the cooperative principles of "democratic member control" and "concern for community."[16]

While posing challenges such as these, Francis turns his listeners' attention to the past. As so often in Catholic discourse, tradition serves as a source of possibilities for the future. During the same 2013 address in which he recalled his father's endorsement of cooperation, Francis said:

> I hope that all of you who are committed to cooperative reform, will keep alive the memory of their origins. The cooperative forms established by Catholics such as the implementation of *Rerum Novarum* bear witness to the power of faith, which today as then is capable of inspiring concrete action to respond to the needs of our people.

The reference to Pope Leo XIII's 1891 encyclical serves as a placeholder and a reminder: We are not stuck with solely the resources of the present. It's an allusion to the long history of cooperative experience in Christian faith and Catholic cultures. To hear the resonance of Francis's enthusiasm for cooperative enterprise, this often-neglected history merits recapitulation, even if it must be rough, brief, and cherry-picked from a vast and diverse tradition irreducible to any one type of economic advice.

"ALL THINGS IN COMMON"

The biblical narrative for the birth of the Christian church describes a social order that bears at least a family resemblance to cooperative economics. Twice in the Book of Acts' account of the days following Pentecost, we read about Jesus's followers pooling property. Here is the first time, in chapter two:

> Awe came upon everyone, and many wonders and signs were done through the apostles. All who believed were together and had all things in common; they would

sell their property and possessions and divide them among all according to each one's need.[17]

Echoes of this language reappear again in the fourth chapter, in which, two verses after another experience of "signs and wonders," we read that "the community of believers was of one heart and mind, and no one claimed that any of his possessions was his own, but they had everything in common."[18] The subsequent chapter then tells the story of Ananias and Sapphira, who die sudden deaths after attempting to withhold from the community part of their earnings from a sale of land. When their story ends, again, "Many signs and wonders were done among the people at the hands of the apostles."[19] Evidently there is a link between the experience of divine activity in the world and the sharing of property among members of the Christian community; dishonest dealing in this arrangement has dire consequences. The lesson of Ananias and Sapphira holds a place of honor at St. Peter's Basilica in Rome, where a depiction of it appears on the canvas over the Altar of the Lie.

A chapter later, the apostolic commune experiences more growing pains. The original apostles find the task of resource management beyond their ken, evidenced by a realization that the needs of widows are being neglected. They ask the community to select a board of seven trusted representatives to carry out the distributions.[20] Their cooperation, as in modern cooperations of any significant scale, was not opposed to hierarchy – but the hierarchy is appropriate and accountable.

This feature of the Bible, like so many others, keeps coming back to haunt Christendom, despite its imperial pretensions. The economic spirit of Acts reappeared with the monastic movements that arose and flourished as Christianity took institutional form in the later centuries of the Roman Empire. Citing Acts, the early-fifth-century Rule of St. Augustine instructs monastics, "Call nothing your own, but let everything be yours in common."[21] About a century later, Benedict of Nursia went further in his rule, stipulating, "As often as anything important is to be done in the monastery, the abbot shall call the whole community together," discussing the matter with everyone before making a decision.[22] Benedict's rule prescribes election of the abbot by the community and enjoins the community to support itself through shared businesses. Both rules emphasize obedience to the abbot or abbess over democratic deliberation. But they also enjoin an egalitarian spirit and a collaborative economy.

The importance of consultative governance recurred again during the thirteenth-century mendicant movement. Clare of Assisi, Francis of Assisi's friend and colleague, enshrined in her rule for Franciscan sisters a measure of counter-cultural self-governance. The draft of the rule that Pope Innocent IV proposed for her order required that the sisters' elected abbess must gain the approval of the male friars' minister general, a provision Clare struck from the final version. She also added the practice of a weekly chapter meeting – which apparently the pope deemed unnecessary – in which the sisters gather to confess their offenses and discuss "the welfare and good of the monastery." She stressed the inclusion of all community members in this, noting, "the Lord often reveals what is best to the lesser [among us]."[23]

Meanwhile, in the secular economy, a parallel model of cooperative-like enterprise appeared with the rise of the guild system. Guilds often took explicitly Christian form, providing members with coverage for religious expenses such as burial (as well as justifying the economic exclusion of Jews, women, and others whose participation might challenge the allegedly Christian power arrangements of the time).[24] While the guilds were highly regimented organizations, like monasteries, and demanded far less wealth sharing, they did serve as a means for workers to participate in co-managing their economic lives, and to do so through a shared commitment to Christian solidarity.

Obviously medieval society was far from the absolute communism of the post-Pentecost *ekklesia*. Church doctrine recognized this tension and sought to reconcile it. Gratian's twelfth-century *Decretum*, the compendium of canon law that steered governance in the church for 800 years, held that "all things are common to all men." By the lights of divine natural law, at least, private property is an aberration, though, under the conditions of fallen human society, it's a necessary, temporary arrangement.[25] This doctrine has come to be called, including in the current catechism,[26] "the universal destination of goods." Thomas Aquinas affirmed it, stressing that "man ought to possess external things, not as his own, but as common, so that, to wit, he is ready to communicate them to others in their need."[27] While these commitments do not predetermine the ownership structure of businesses among Christians, they do seem to encourage practicing private property in a way that reflects its provisional quality and the communal reality beneath it. There were and are many ways of doing so; in the twelfth century, for instance, the (female) Beguines and (male) Beghards maintained private enterprise and property while sharing real estate.

As during the early monastic and mendicant movements, practices of shared ownership have tended to feature prominently whenever reformers seek to challenge the moral authority of the church of their day and return to the core of the Gospel as they understand it. This was no less the case during the Protestant Reformation. While the Reformation helped ensure the rise of capitalist ownership structures along individualistic lines, it also occasioned rhetorical and institutional innovations oriented toward cooperative economics.[28]

What has come to be called the "Radical Reformation" offers plentiful examples. Luther's militant rival Thomas Müntzer is said to have cried out, while being tortured to death in 1525, "*omnia sunt communia*," or "all things are common"; more peaceable Anabaptist groups, meanwhile, were more lasting in their efforts to practice that biblical principle in everyday life and business – down to the Mennonites, Amish, and Hutterites still active today. In the midst of the English Civil War, the Diggers, under the eloquent leadership of Gerrard Winstanley, set out to win (by practicing, on a small scale, before their defeat) a social order based on common ownership of land, against both the receding feudal order and the emerging capitalist one.[29] The Calvinist variant of the communal impulse found expression among the Puritans who settled in New England; though their initial experiment in shared farmland and produce was short-lived, their legacy has lived on in the region's town meetings and the congregationalist governance structure that still holds sway in much of US Protestantism.

None of these examples, Catholic or Protestant, exactly corresponds with cooperative enterprise in the modern sense, and in certain respects they expressly challenge cooperative principles – for instance, in monasteries, the preference for poverty over ownership and obedience over autonomy. Nevertheless, such precursors indicate that, when the highest authorities of the Catholic Church began to take an interest in cooperatives as a response to the crises of modernity, they acted in continuity with longstanding traditions that preceded them. And in *Rerum Novarum*, the 1891 encyclical to which Francis alluded, Leo XIII describes the challenge of industrial labor relations as a breach in that continuity; "the ancient workingmen's guilds were abolished in the last century," he worried, "and no other protective organization took their place."[30]

APOSTOLIC VOICES

A few years ago, on the heels of a conference in Quito, Ecuador, I joined a handful of cooperative business experts from various countries on an excursion to Salinas de Guaranda, a nearby mountain village known for its network of cooperative enterprises. What we found there was remarkable. We visited factories making cheese, chocolate, and textiles, as well as a hostel, a dairy, and a credit union, all in various ways owned by and accountable to Salinas's residents through a bewildering organizational structure. The village, too, had become an engine for development throughout the region, spurring offshoots and replicas. Our hosts allowed me to join the weekly town meeting in an upper room next to the church on the main square, under the concrete cross mounted on the hill above; the gathering doubled, I soon discovered, as a prayer meeting. The whole system arose in the 1970s from the instigation of an Italian priest.[31]

Salinas is in some respects an exception – to the usual economic hopelessness of remote mountain towns, to the usual craving for rescue by multinational corporations – and in some respects not. Go on enough tours of impressive co-ops, and the story of Catholic influence repeats itself. The first credit union in the United States, St. Mary's Bank, was born of a New Hampshire parish in 1908 with the help of Alphonse Desjardins, founder of what would become Quebec's vast financial federation. Catholic Relief Services and various missionary orders have helped establish producer cooperatives around the world that enable poor farmers to obtain fairer prices in globalizing markets. The largest worker cooperative on record is Basque Country's Mondragon Corporation, which employs over 70,000 people in high-tech industry, finance, and retail; like Salinas, it began with an enterprising priest.

This pattern may be at least partly attributable to the prehistory already outlined. But a legacy of monasteries, guilds, and biblical passages also seems inadequate to explaining the leadership that Catholics have exercised in the specific domain of modern cooperative enterprise. For that, Francis's nod to *Rerum Novarum* is instructive.

Cooperative enterprise arose in tandem with the rise of the industrial capitalism, although the length and breadth of this history is rarely known or told, even among cooperators themselves.[32] As capital-intensive industry engulfed small trades and crafts, workers repeatedly sought arrangements in which they could retain ownership of the means of production and the revenue from their labor. Though its origins go back at least several decades earlier, the modern cooperative movement is often traced to the 1844 founding agreements for a cooperative store in Rochdale, England, which would become an influential model elsewhere. These principles enshrined one-member-one-vote democracy, the distribution of surpluses to members, transparent accounting, and the promotion of education. Religious influences were certainly at work in Rochdale, but the orientation was ecumenical, not ecclesiastical; when the International Co-operative Alliance distilled the lessons of the Rochdale store in 1937, the first was "open membership" and the fifth was "political and religious neutrality."[33] To the Rochdale Society of Equitable Pioneers, as the store's founders called themselves, the ancient and hierarchical Catholic Church probably seemed remote from the new cooperative economic order they envisioned.

It was only later in the nineteenth century that the foundations began to form for a specifically Catholic cooperativism. When Leo XIII promulgated *Rerum Novarum* (or, *Rights and Duties of Capital and Labor*) in 1891, he appears to have understood his church as standing between a materialist-socialist rock and a robber-baron-capitalist hard place. Decades of revolution and class conflict had shaken Europe, cracking many of the church's familiar secular buttresses. He laments "the enormous fortunes of some few individuals, and the utter poverty of the masses" that industrial capitalism had brought about, but he cannot accept the socialists' alternative, which would "rob the lawful possessor, distort the functions of the State, and create utter confusion in the community."[34] His third-way solution, rather than accumulating power in either capital or the state, is to double down on private property and provide for its more widespread distribution: "The law, therefore, should favor ownership, and its policy should be to induce as many as possible of the people to become owners."[35]

To accomplish this, too, Leo affirms the need for Christian workers' organizations to demand fair wages and facilitate mutual aid. But this should take harmonious, Christian form, eschewing acrimonious strikes and riots, for "capital cannot do without labor, nor labor without capital."[36] Even while asserting the priority of property, he retains the spirit of the universal destination of goods, by which property should be used in the service of the commonweal. The encyclical doesn't include an explicit embrace of Rochdale-style cooperation, while nevertheless providing the basis for its subsequent adoption – pointing away from state socialism and toward forms of economic life that enable more collaborative labor relations. The solution Leo offers workers is not to expropriate the means of production and deliver them to a high and mighty state, but to seek more equitable ownership over them for themselves.

Exactly forty years later, in the 1931 encyclical *Quadragesimo Anno*, Pope Pius XI heralds at length the reception history of *Rerum Novarum*: "The Apostolic voice did not thunder forth in vain," he boasts.[37] He speaks of the various labor organizations and study circles and other communal experiments that emerged from his predecessor's provocation. Even as the Great Depression swelled to full force, his tone is of momentum and optimism.

In England, Leo XIII's outline for more widely distributed ownership had begun to crystallize into a *sui generis* political philosophy. Cardinal Henry Manning of Westminster, one of the major contributors to *Rerum Novarum*, had a young acolyte in Hillaire Belloc, whose mother's conversion to Catholicism Manning had steered. Belloc became a public intellectual in his own right, and much of his career could be summarized as an attempt to interpret and defend the vision of *Rerum Novarum*. In the process he introduced the encyclical and its implications to Cecil and Gilbert Chesterton, which aided in converting the brothers to Catholicism and away from statist socialism. The three called their synthesis "distributism." Together, in bellicose polemics and debates, they described a distributist system of family businesses and parish associations, although they could rarely adequately satisfy questions of how to get there. They had a tendency to level accusations at high finance in anti-Jewish terms and cheer along the ascendant fascists. From time to time, they mentioned cooperatives.[38]

Pius XI, in *Quadragesimo Anno*, also sought to further and systematize what *Rerum Novarum* began. He put a name to the logic of subsidiarity that animates *Rerum Novarum*, the principle that governance should remain as local as possible. (A theme that Catholic social teaching often pairs with subsidiarity – that of solidarity – is implicit in the document but not explicitly stated.) Although affirming the basic justice of wage-based employment, he gestures toward the kind of shared equity that is so central to cooperative enterprise, recommending that

> the work-contract be somewhat modified by a partnership-contract, as is already being done in various ways and with no small advantage to workers and owners. Workers and other employees thus become sharers in ownership or management or participate in some fashion in the profits received.[39]

The encyclical was drafted chiefly by a German Jesuit who later stated his intent as "progressive, liberal, definitely democratic, against individualism and against statism," but the final text and its reception turned out to be more ideologically ambiguous.[40] Late in the drafting process, Pius XI added a passage that addressed, though not by name, the fascist system then emerging under Benito Mussolini in Italy: a tripartist collaboration among capital, labor, and the state. He approved of how such a system could align competing interests of industry toward the common good. But the pope's comments expressed concern that the state was overreaching its bounds and dominating the other parties; "instead of confining itself as it ought to the furnishing of necessary and adequate assistance," he worried, the regime "is substituting itself for free activity."[41] He recommended, as an antidote, the kinds of lay Catholic Action organizations that were then developing cooperatives, among other things, and which Mussolini had been suppressing. Mussolini complained about these comments; Pius XI claimed they were merely a "benevolent nod." Many since have read the encyclical's critique as dangerously close to an endorsement.

Between Pius XI's prewar involvements with Catholic dictators and the English distributists' racist fits, the ideological foundations of Catholic cooperativism make for a checkered pedigree. What's more, the doctrines of neither the popes nor the distributists provided a thorough blueprint for the harmonious society of equitable ownership and subsidiaristic governance to which the authors aspired. It was in places like Salinas, in the field of practice, that a distinctly Catholic and remarkably effective approach to cooperative enterprise would take shape.

COOPERATION THROUGH FORMATION

After a talk I gave in Melbourne, Australia, in mid-2016, a friend introduced me to an elderly man who'd been in the audience.[42] The man had something he wanted to say. He came close to me and began to speak, from which I recollect only one crucial word: *formation*.

The man's name was Race Mathews. And despite his reference to a theological concept, he confessed to being neither a Catholic nor a believer. For most of his career, before retiring to study the prehistory of the modern cooperative tradition, he was a politician and official in Australia's Labor Party. The importance of formation was his latest discovery.

Formation is a word now used frequently in Catholic discourse, though its distinct Catholic usage is rarely defined outright. It refers to a Christian's ongoing conversion to Christianity – through prayer, study, and experience. The kind of formation one receives depends on one's outside influences and one's interior choices, and it determines what kind of Christian one becomes. Economic life is part of a person's formation, too.

Since the 1980s Mathews has made a series of visits to the Mondragon Corporation. This remarkable network of worker co-ops emerged under the shadow of Franco in the 1950s through the guidance of a local priest, José María Arizmendiarrieta – Arizmendi for short. It's a system of factories, schools, banks, retailers, and more, all owned and governed by people who work in them. It's a beacon of possibility, the world over, that democratic business can work at large scale, though it has yet to be outdone or replicated.

Mathews' 2009 book *Jobs of Our Own* traces Arizmendi's precursors, including *Rerum Novarum* and the British distributists. In Mondragon, Mathews saw a practical manifestation of the ideas earlier Catholics had only alluded to, dubbing it "evolved distributism." But only after publishing the book, while studying the Catholic Action and Young Christian Workers movements that influenced Arizmendi, did Mathews zero in on the concept of formation.

Mondragon, he realized, is a monument not only to a particular way of doing business but to a vision for forming the souls who partake in it. Before Mondragon opened for business in 1956, Arizmendi started a school where he and his students developed their plans together over the course of a decade. They let diverse influences form them, religious and otherwise. Following the teachings of Young Christian Workers founder Cardinal Joseph Cardijn – teachings that also influenced Jorge Bergoglio – they practiced the method of "see, judge, act." They tested their ideas relentlessly and creatively through practice, and then adjusted those ideas accordingly. "Hand in hand, of one mind, renewed, united in work, through work, in our small land we shall create a more human environment for everyone," Arizmendi wrote just days before his death in 1976. "Everyone shall simply work for the benefit of everyone else, and we shall have to behave differently in the way we work."[43]

The centrality of education for Mondragon is hard to overstate. "It has been said that cooperativism is an economic movement that uses the methods of education," Arizmendi once wrote. "This definition can also be modified to affirm that cooperativism is an educational movement that uses the methods of economics."[44] The reasons for Mondragon's success are complex enough to be infinitely open to interpretation, but among them is surely the view of participants' spiritual development as an end in itself, rather than simply a means toward enacting an abstract economic system.

This emphasis points to a further, vital influence for Arizmendi and his followers, alongside *Rerum Novarum* economics. The priest's teachings frequently echoed the philosophy of personalism articulated by Catholic thinkers like Jacques Maritain and Emmanuel Mounier, in which recognizing and cultivating the God-bestowed dignity of human persons constitutes the heart of the Gospel. Formation, in this light, is no rote indoctrination; it is an invitation to divine freedom and flourishing.

Arizmendi's personalism also did much to escape the reactionary tendencies of earlier, more theoretical Catholic cooperativism. He taught, at least later in life, a commitment to gender equality. "To advance in life without the collaboration of women," he wrote, "is to attempt to run with only one foot."[45] And emerging as Mondragon did in quiet defiance of the Franco regime's repressions of the Basques, it belied its predecessors' fascist flirtations.

While Mondragon was uniquely successful, this cooperative expression of Catholic social teaching was far from unique. Across the Atlantic, a separate but very much parallel manifestation of Catholic cooperativism appeared.

Antigonish, Nova Scotia, like Mondragón, was an impoverished place far from the capitals of culture and industry – at the "periphery," to use a favorite word of Pope Francis's. And again, what would become the Antigonish movement began with a priest and a school. Fr. Jimmy Tompkins, and then Fr. Moses Coady, starting in the mid-1920s, used a Jesuit university's extension program as a vehicle for reaching working people with cooperative ideas and study groups, learning together about the Desjardins credit union in Quebec and the Rochdale model. With the onset of the Great Depression, the study turned into organizing. By 1932, they'd set up around 200 local organizations with around 10,000 members. This spurred a flourishing of co-ops, credit unions, and study groups throughout the region.[46]

As Arizmendi would do in decades to come, Tompkins and Coates brought the spirit of *Rerum Novarum* into a patient, practical process of education and formation. Alongside encyclicals, they drew freely from secular experiments and the know-how of the people they worked with. The Antigonish movement gained enough international visibility by 1938 to merit a letter from the Vatican under Pius XI, signed by Cardinal Eugenio Pacelli, who would become Pope Pius XII the following year; "May the work grow and flourish," it said.[47] The late 1930s, as it happened, would mark the movement's peak, but the Antigonish model would help spur a broader Catholic cooperative movement that would continue to mature and proliferate.

One can try one's best to assemble a litany of Catholic cooperative efforts in the past century, but in the process one becomes all the more aware of coming up short. The Confederation of Italian Cooperatives, founded in 1919, led to a flourishing of specifically Catholic co-ops throughout the country, alongside those associated with liberals and communists, especially in the Emilia-Romagna region. In the Southern United States, the African-American Spiritan priest Fr. Albert J. McKnight became one of the architects of a cooperative system that enabled thousands of black farmers to keep their land and make a living – despite a racist social order that their co-ops would help embolden them to resist in the Civil Rights movement. President Jimmy Carter appointed McKnight to the founding board of the National Cooperative Bank in 1979.[48] Meanwhile in New York, the lay-edited magazine *Jubilee*, active between 1953 and 1967, organized itself as a consumer cooperative, with shares owned by subscribers; this model took inspiration from Clare Hutchet Bishop's *All Things in Common*, a lyrical dispatch from among the cooperative industries that appeared throughout France after World War II. Around the world, Catholic mission agencies have supported cooperatives among poor farmers and craftspeople, which enable them to bring their products into the global market without relying on multinational corporations.

All along, the pedagogical aspect of cooperation – the role in formation – has been of particular interest to Catholics. This benefit was evident enough by 1916, for instance, that the prominent US Catholic economist Monsignor John A. Ryan devoted a chapter to co-ops in his book *Distributive Justice*. In addition to concluding that cooperation "always makes directly for a better distribution of wealth," he writes:

> Co-operation has, in the second place, a very great educational value. It enables and induces the weaker members of economic society to combine and utilise energies and resources that would otherwise remain unused and undeveloped; and it greatly stimulates and fosters initiative, self confidence, self restraint, self government, and the capacity for democracy. In other words, it vastly increases the development and efficiency of the individual ... Its fundamental and perhaps chief merit is that it does provide the mechanism and the atmosphere for a greater development of the altruistic spirit than is possible under any other economic system that has ever been tried or devised.[49]

Ryan's ideas would go on to influence the shape of the New Deal, whose programs led to a massive expansion of agricultural and electric-utility co-ops across the United States. In 1919 he also drafted the US bishops' *Program of Social Reconstruction*, which regards wage-based employment as inadequate and commends cooperative ownership models instead; the bishops' landmark 1986 statement *Economic Justice for All* quotes from those remarks before again encouraging experimentation in cooperative enterprise, citing Mondragon.[50]

While many of the most full-throated co-op advocates nowadays are self-described progressives, cooperativism and related discourses continue to cross left-right political lines much as they did a century ago among the British distributists. Legislators representing conservative rural districts are among the first to champion electric co-ops and credit unions. Catholics have been notably vocal. Both Republican Jeb Bush and Democrat Tim Kaine, for instance, have advocated easing financial regulations on credit unions and community banks; Republican congressman Paul Ryan has flummoxed progressive Catholics by defending his proposals for welfare reform with the language of subsidiarity.

In my reporting on the young people who are transforming the cooperative movement today, I frequently encounter non-Catholic cooperators who take it for granted – without necessarily knowing why – that Catholic organizations have been steady and quiet supporters of the work they do. The US bishops' Catholic Campaign for Human Development, for instance, has invested in the fledgling US Federation of Worker Cooperatives, as well as Cooperation Texas, a regional co-op developer, and individual co-ops like Opportunity Threads, a worker cooperative "eco-textile" plant in North Carolina.[51] Catholic Relief Services is an "interfaith partner" of Equal Exchange, a cooperative coffee company, encouraging parishes across the country to serve the company's fair-trade products. On a smaller scale, the Jesuit parish of St. Robert Bellarmine in Cincinnati, Ohio, helped open Community Blend, a worker-owned cafe, in 2014.[52] Cincinnati's Xavier University, to which the parish is connected, held a conference in April 2016 on "The Cooperative Economy: Building a Sustainable Future."

Many of these efforts take place without publicity, and often Catholics themselves aren't aware of the leadership role their co-religionists have taken and are taking in creating cooperative economic alternatives. One Jesuit professor, the assistant to the University of San Francisco's president, lamented in 2000, "Considering the number of business schools and law schools in U.S. Catholic universities, the lack of creative thinking on cooperatives is surprising."[53] Yet the practical achievements of Catholic cooperativism have secured the notice of popes, at least, on a regular basis.

"It is not possible to give a concise definition of the kind of economic structure which is most consonant with man's dignity and best calculated to develop in him a sense of responsibility," John XXIII's *Mater et Magistra* of 1961 admits.[54] Yet the document commends Pius XII's encouragement of small producers banding together in cooperatives. Especially in agricultural life, it concludes, "We therefore desire here to express Our satisfaction with those sons of Ours the world over who are actively engaged in co-operatives."[55]

A 1998 address to credit unionists by John Paul II began with some credit-taking. "The institution you represent," he said, "draws its inspiration precisely from the Church's fruitful Magisterium and is one of its most significant concrete achievements."[56] He cited *Rerum Novarum*'s influence in spurring the co-op movement and warned, "These values must be safeguarded in a 'market' that is always in danger of forgetting that" – here referencing his own encyclical *Sollicitudo Rei Socialis*, together with the ancient universal destination of goods – "the goods of creation are meant for all."

Benedict XVI also discerned agreeable resonance between the cooperative model and Catholic teaching. In his December 2011 address to the Confederation of Italian Cooperatives, he said of the model:

> It is a practical expression of complementarity and subsidiarity which the Church's social doctrine has always encouraged between the person and the State; it is the balance between the safeguard of the rights of the individual and the promotion of the common good in the effort to develop a local economy that responds more and more to the needs of the group.[57]

He went on to encourage the Italian co-ops to keep Christian spirituality at the center of their work – living out an economy "of giving freely, of responsibility, of promoting a responsible and moderate consumption."

The following year, Benedict XVI shared a message to the United Nations on World Food Day, whose theme that year (as part of the UN International Year of the Cooperative) centered on agricultural cooperatives. Again he raised the theme of co-ops as a mechanism for forming participants in the "human, social, cultural and moral development" that the church seeks to engender; he also celebrated the leadership role women often take. "It is indispensable that national and international authorities provide the necessary legislation and financing" to support rural cooperation, he concluded.

One could go on. Suffice it to say that, by the second half of the twentieth century, popes had more to go on than *Rerum Novarum*'s broad-stroke recommendations for more equitably distributed ownership. There was enough actually existing Catholic cooperation to demonstrate the model's prowess both for establishing fair economic conditions and for forming the souls of participants in conformity with Catholic teaching. Already, too, Catholics had become leaders in the cooperative movement the world over, from famous cases in point such as Mondragon to the countless, unsung co-ops that church relief organizations have supported in precarious communities.

Pope Francis's own plentiful statements about cooperative enterprise stand in continuity with those of his predecessors; their existence comes as no surprise. Yet Francis also demonstrates a special tenderness for cooperation, as well as an especially sophisticated understanding of its prospects and pitfalls. He wields the co-op alternative as part of his barbed critique of the capitalist order's "idolatrous economy," while encouraging cooperators to resist the temptation to compromise too much. Rather than uplifting co-ops only in a specific sector, also, he recognizes that they can take many forms, in many industries, and can serve as a means for addressing a variety of social ills. In particular, and particularly in *Laudato Si'*, he identifies the resonance between cooperative enterprise and an aspect of the Christian calling that his predecessors didn't similarly appreciate: stewardship of God's creation.

"SOCIAL WORSHIP"

America magazine, the national Jesuit weekly,[58] has a front-of-the-book section of short editorials opining on issues of the day. The October 21, 1944, issue includes one such entry, entitled "Catholics and Cooperatives," alongside a call to "Recognize DeGaulle." It opens by paying tribute to "Iowa in October" – "a lovely country" of "rolling prairie" and "glowing colors" – before turning to the human action. In the town of Westphalia, the unnamed author describes a parish community celebrating the Rochdale Centennial with apple picking, cow milking, and a solemn mass, all under the aegis of the Catholic Cooperative Committee, which supports the parish's several cooperatives. A bishop is quoted speaking approvingly. The author takes it for granted that this is but one place where similar scenes are playing out across the country, for the cooperative model "has long since passed the experimental stage." This is "a parish," we're told, "which draws the light and strength for its economic life from the Catholic ideal of truly social worship."

The ease and normalcy with which the editorial speaks of Catholic cooperation is no longer with many of us. In the years after World War II, the US cooperative movement fell victim to postwar prosperity and Red-baiting repression. Many of those cooperatives that have remained in force, particularly among farmers and rural utilities, have become rigid to the point that their members are not even aware of being members of a cooperative. Co-ops that do claim their identity with pride, meanwhile, often seem to serve marginal countercultures and thereby lack appeal for the mainstream. The everydayness of the editorial reminds us that this was not always the case.

At the same time, the "social worship" so eloquently named there prefigures a calling very much of our moment. In the juxtaposition between the natural world – that "lovely country" – and the harmonious economy operating therein, we should recognize an example of the "integral ecology" that Francis proposes in *Laudato Si'*. Unlike the scientistic or animistic visions that have long motivated secular environmentalism, integral ecology aspires to a form of stewardship that integrates natural science, social justice, and authentic spirituality without sacrificing any of them to the others – "an approach to ecology," Francis writes, "which respects our unique place as human beings in this world and our relationship to our surroundings" (LS 15).

By now it should be evident that Catholics have long envisioned and practiced cooperation as an integration of structural social justice and day-to-day spirituality. But where does ecology fit?

The encyclical's subtitle, *On Care for Our Common Home*, alludes to the universal destination of goods, which lends new meaning to Genesis in an age of planetary crisis. "All of us can cooperate as instruments of God for the care of creation," he writes; elsewhere, he speaks of a God "who wishes to work with us and who counts on our cooperation" (LS 14, 80). Francis challenges readers to establish an economic order that makes a priority of long-term sustainability rather than short-term extractive gain – an "economic ecology," as he puts it (LS 141). And there is every reason to surmise that cooperative enterprise is a sensible way to begin practicing this.

The pioneering work of political scientist (and Nobel laureate in economics) Elinor Ostrom suggests how forms of local self-governance and mutual investment have resulted in efficient, ecologically sustainable outcomes.[59] In countries like Germany and Denmark that have been leading the turn from fossil fuels, cooperatives have been a formidable mechanism for financing and governing renewable energy systems; even in the United States, green-energy advocates see existing rural electric co-ops as potential leaders in the transition because they're more responsible to the holistic needs of communities they serve than to investors' cravings for short-term profits.[60] Francis refers to such prospects in *Laudato Si'*:

> In some places, cooperatives are being developed to exploit renewable sources of energy which ensure local self-sufficiency and even the sale of surplus energy. This simple example shows that, while the existing world order proves powerless to assume its responsibilities, local individuals and groups can make a real difference. They are able to instill a greater sense of responsibility, a strong sense of community, a readiness to protect others, a spirit of creativity and a deep love for the land. They are also concerned about what they will eventually leave to their children and grandchildren. These values are deeply rooted in indigenous peoples. (LS 179)

Here we see that Francis views local cooperation as a source of innovation, not just as a means of hearkening back to a slower, simpler existence. Earlier he writes, "We have the freedom needed to limit and direct technology" – but this is not a call for mere retrenchment. "We can put it at the service of another type of progress," he goes on, "one which is healthier, more human, more social, more integral" (LS 112). That precedes the first of the encyclical's three explicit references to cooperative enterprise (though the last quoted here):

> Liberation from the dominant technocratic paradigm does in fact happen sometimes, for example, when cooperatives of small producers adopt less polluting means of production, and opt for a non-consumerist model of life, recreation and community.

This is not liberation from technology as a whole, but from a paradigm in which the apparent needs of profit-oriented technology take precedence over social, spiritual, and ecological flourishing. Likewise, Fr. Arizmendi insisted that cooperatives must be technologically dynamic and well capitalized enough to innovate; to this end, he founded Mondragon Corporation with students at a technical school, and ever since it has maintained units devoted to R&D. He took the imperative of innovation for granted, though not as an end in itself. "Tools and machines need to be continuously renewed," Arizmendi wrote, "but above all there has to be a renewal in the mentality of human beings."[61]

Francis regards cooperative enterprise as a proven method for putting economic activity under the control of stakeholders who have more at stake than investors playing with money; these are the workers, the customers, the producers, the neighbors, the people who drink the nearby water and breathe the same air. Cooperation is for him, therefore, a cause for hope. It is not the ultimate goal or a one-size-fits-all solution to everything; he points also to international agreements, policy tweaks, changes in personal habits, and more. But cooperation does provide a reason to believe that the ambition of integral ecology is not beyond reach. If one is familiar with the cooperative legacy, the encyclical's central doctrine is more likely to seem practically plausible and intellectually satisfying. If not, one may be more likely to view the doctrine as dark and foreboding – a lost cause, a charge toward collective martyrdom. But that is not the intent. Bound up in the call for integral ecology is a call for a cooperative revival.

To begin with, it is a call for those who claim allegiance to Catholic faith and its social teaching to reacquaint themselves with the theory and practice of cooperative enterprise. In more than a decade since becoming a Catholic I have neither seen nor heard of an extant Catholic Cooperative Committee at a parish; I've never heard a priest recommend economic cooperation from the pulpit or a leading Catholic intellectual mention it in the mass media – at least in the United States. When I visited Rome to learn more about the backdrop of the encyclical, I encountered economists from the old Italian "civil economy" tradition such as Stefano Zamagni and Leonardo Becchetti, who spoke of cooperatives as a natural part of a healthy society. It's no accident that Zamagni has been an influential shaper of recent social encyclicals.

There are many opportunities for Catholics to reclaim their leadership role in the cooperative movement. Catholic schools and hospitals can act as "anchor institutions," using the money they expect to spend on essentials like food, laundry services, construction, and electricity to finance worker-owned green businesses.[62] Parishes and dioceses can operate credit unions equipped to supplant abusive payday lenders and provide financing for rooftop solar panels. Networks of such cooperatives, connected through existing church institutions, can share local best practices globally and invest in technology that meets their needs more efficiently. Perhaps before all else, Catholic communities can form study groups to reflect on the economic formation at work in their lives and how they might align their economies more closely with an integral ecology. As Francis recommends at the outset of the encyclical, they can do all these things with non-Catholic partners in pursuit of the common good.

For all the empowerment that economic democracy might offer, however, for Francis it renders no substitute for humility. "It is important that we not think of ourselves as self-sufficient," he has counseled elsewhere.[63] The pope nevertheless allows himself to express hope in *Laudato Si'* for implementing integral ecology – with an exclamation mark, no less – in part because he knows the accomplishments of alternatives like the cooperative model, as well as his predecessors' role in helping to motivate it. If *Rerum Novarum* eventually led to Mondragon, perhaps *Laudato Si'* can inspire a green-energy revolution whose benefits accrue more widely, and more accountably, than those of the oil conglomerates. If workplaces operated more like a village dairy cooperative in the Andes, perhaps work would start to feel more like social worship. If an economy can regard its participants as co-creators, maybe they'll rediscover their role as stewards of creation. In the meantime, much can be done.

Notes

* This chapter considerably expands on an earlier publication of mine, "How Pope Francis Is Reviving Radical Catholic Economics," The Nation (September 9, 2015), and it was the basis for parts of my book, *Everything for Everyone: The Radical Tradition That Is Shaping the Next Economy* (Nation Books, 2018). It has benefited from feedback by Tim Huegerich, Mac Johnson, and Frank Pasquale.

1. I attempted to put the question of mood out into the open with "The Mood of 'Laudato Si'': A Reply," *America* (June 22, 2015), which in turn garnered a response by Matthew Schmitz: "How 'Laudato Si'' Praises God – and Smashes False Idols," *America*, June 23, 2015: www.americamagazine.org/content/all-things/how-laudato-si-praises-god-and-smashes-false-idols
2. Michael Sean Winters, "*Laudato Si'* Arrives," *National Catholic Reporter*, June 18, 2015.
3. R. R. Reno, "The Weakness of *Laudato Si'*," *First Things*, July 1, 2015: www.firstthings.com/web-exclusives/2015/07/the-weakness-of-laudato-si; Matthew Schmitz, "Pope Francis Wants to Roll Back

Progress. Is the World Ready?" *Washington Post*, June 18, 2015; Ross Douthat, "Pope Francis' Call to Action Goes Beyond the Environment," *New York Times*, June 20, 2015.

4. Ibid., § 180.
5. I use terms like "cooperation," "cooperativism," and "cooperative enterprise" to refer to the same phenomenon, though each carries a somewhat different emphasis.
6. See the seven principles presently held in the global cooperative movement at "Co-operative Identity, Values and Principles," International Co-operative Alliance: https://is.gd/tohMNl
7. Dave Grace and Associates, *Measuring the Size and Scope of the Cooperative Economy: Results of the 2014 Global Census on Cooperatives* (United Nations Secretariat Department of Economic and Social Affairs Division for Social Policy and Development, 2014).
8. Austen Ivereigh, "Gaucho Cardinal" in *The Great Reformer: Francis and the Making of a Radical Pope* (New York, NY: Henry Holt, 2014), p. 253.
9. Ivereigh, "Man for Others," in *The Great Reformer*, p. 302.
10. On June 19, 2015, with translation assistance from Caracol Interpreters Cooperative.
11. Pope Francis, "Video message for the Third Festival of the Social Doctrine" (November 21–24, 2013).
12. "Pope Francis: Speech at World Meeting of Popular Movements," Vatican Radio (July 10, 2015): www.news.va/en/news/pope-francis-speech-at-world-meeting-of-popular-mo
13. Ivereigh, "A Bishop Who Smelled of Sheep (1993–2000)" in *The Great Reformer*, p. 210; Pope Francis, *Evangelii Gaudium* (2013).
14. Pope Francis, "Address of His Holiness Pope Francis to Representatives of the Confederation of Italian Cooperatives" (February 28, 2015): https://w2.vatican.va/content/francesco/en/speeches/2015/february/documents/papa-francesco_20150228_confcooperative.html (emphasis in the Vatican transcript).
15. "Pope Francis Encourages Cooperatives to Build Solidarity," Vatican Radio, May 5, 2016: www.news.va/en/news/pope-francis-encourages-cooperatives-to-build-soli
16. Again, see "Co-operative Identity, Values and Principles," International Co-operative Alliance.
17. Acts 2:43–45. This and all subsequent biblical quotations are from the New American Bible: http://usccb.org/bible
18. Acts 4:32.
19. Acts 5:12.
20. Acts 6:1–6.
21. Augustine of Hippo, *The Rule of St. Augustine*, trans. Robert Russell (Villanova, PA: Brothers of the Order of Hermits of Saint Augustine, 1976): https://sourcebooks.fordham.edu/source/ruleaug.html
22. Benedict of Nursia, *RB 1980: The Rule of St. Benedict in English*, trans. Timothy Fry (Collegeville, MN: Liturgical Press, 1981), p. 25. The relevance of the Benedictine rule to modern cooperativism is the subject of Greg MacLeod, "The Monastic System as a Model for the Corporate Global System," *Work as Key to the Social Question* conference on social justice, Vatican City (September 12–15, 2001).
23. Chapter four of the Rule of St. Clare, in *Francis and Clare: The Complete Works*, trans. R. J. Armstrong (New York, NY: Paulist Press, 1988), pp. 215–16 (bracketed words in the original).
24. Steven A. Epstein, *Wage Labor and Guilds in Medieval Europe* (Chapel Hil, NCl: University of North Carolina Press, 1995); Sheilagh Ogilvie, "The Economics of Guilds," *Journal of Economic Perspectives*, 28(4) (2014), 169–92.
25. Robert Harry Inglis Palgrave (ed.), *Dictionary of Political Economy*, vol. one (New York, NY: Macmillan and Co., 1915), p. 212.
26. *Catechism of the Catholic Church*, 2nd edn. (Vatican City: Liberia Editrice Vaticana, 1993), part 3, section 2, chapter 2, article 7.I.
27. Thomas Aquinas, "Is It Lawful for a Man to Possess Something as His Own?" in *Summa Theologica*, trans. Fathers of the English Dominican Province (Denver, CO: New Advent, 2008 [1920]).
28. Cf. Max Weber, *The Protestant Ethic and the Spirit of Capitalism*, etc. A valuable, though more Protestant perspective, on this story of Christian cooperation can be found in Andrew McLeod, *Holy Cooperation! Building Graceful Economies* (Eugene, OR: Cascade Books, 2009).

29. Gerrard Winstanley, *A Common Treasury* (New York, NY: Verso, 2011).
30. Pope Leo XIII, *Rerum Novarum* (1891), 3: http://w2.vatican.va/content/leo-xiii/en/encyclicals/documents/hf_l-xiii_enc_15051891_rerum-novarum.html
31. Andrea Gagliarducci, "The Man Who Put *Laudato Si'* into Practice in Ecuador – 40 Years Ago," *Catholic News Agency* (July 8, 2015).
32. There are many early US examples, for instance, in John Curl, *For All the People: Uncovering the Hidden History of Cooperation, Cooperative Movements, and Communalism in America*, 2nd edn. (Oakland, CA: PM Press, 2012).
33. *The Present Application of the Rochdale Principles* (London: International Co-operative Alliance, 1964).
34. Leo XIII, *Rerum Novarum*, paras 1 and 4.
35. Ibid., para. 46.
36. Ibid., para. 19.
37. Pope Pius XI, *Quadragesimo Anno* (1931), para. 12: http://w2.vatican.va/content/pius-xi/en/encyclicals/documents/hf_p-xi_enc_19310515_quadragesimo-anno.html
38. For this account and much else, I am indebted to Race Mathews, *Jobs of Our Own: Building a Stakeholder Society: Alternatives to the Market and the State* (Sydney: Pluto Press, 2009) and Jay P. Corrin, *Catholic Intellectuals and the Challenge of Democracy* (Notre Dame, IN: University of Notre Dame Press, 2002).
39. Pius XI, *Quadragesimo Anno*, 65.
40. Oswald von Nell-Bruening, "Documentation: The Drafting of *Quadragesimo Anno*," *Crisis Magazine* (February 1, 1985); Paul Misner, "Catholic Labor and Catholic Action: The Italian Context of *Quadragesimo Anno*," *The Catholic Historical Review*, 90(4) (October 2004).
41. Pius XI, *Quadragesimo Anno*, 95.
42. Some readers may recognize passages here adapted from my article "A New Way to Work," *America* (August 15–22, 2016).
43. Quoted in Race Mathews, *Manning's Children: Responses to* Rerum Novarum *in Victoria, 1891 to 1966*, unpublished MCD University of Divinity thesis (March 2014), pp. 94–95: http://repository.divinity.edu.au/1509/1/Mathews_DTheol_Thesis.pdf
44. José María Arizmendiarrieta, *Reflections* (Otalora, 2013), § 213.
45. Arizmendiarrieta, *Reflections*, § 73.
46. Chapters 7 and 8 of Mathews, *Jobs of Our Own*; Jacob Remes, "In Search of 'Saner Minds': Bishop James Morrison and the Origins of the Antigonish Movement," *Acadiensis* 39, no. 1 (2010): 58–82.
47. Mathews, *Jobs of Our Own*, p. 153.
48. "Father Albert McKnight," Cooperative Hall of Fame: http://heroes.coop/archive/father-albert-mcknight/; Albert J. McKnight, Whistling in the Wind: The Autobiography of Fr. Albert J. McKnight (Opelousas, LA: Southern Development Foundation, Inc., 2011).
49. John A. Ryan, "Co-Operation as a Partial Solvent of Capitalism" in Distributive Justice: The Right and Wrong of Our Present Distribution of Wealth (New York, NY: Macmillan, 1916).
50. Administrative Committee of the National Catholic War Council, *Program of Social Reconstruction* (1919); United States Catholic Bishops. *Economic Justice for All: Pastoral Letter on Catholic Social Teaching and the U.S. Economy* (1986).
51. Joseph Purello, "Catholic Campaign for Human Development Funds Textile Cooperative in Morganton," *Catholic News Herald* (December 12, 2012).
52. "Catholic Relief Services Partnership," Equal Exchange: http://equalexchange.coop/catholic-relief-services; Nathan Schneider, "I Was Wrong About Catholic Co-ops," *America*, November 24, 2014: www.americamagazine.org/content/all-things/i-was-wrong-about-catholic-co-ops
53. George E. Schultze, "Workplace Democracy as Church," Peace Review, 12(2) (2000), 235.
54. Pope John XXIII, *Mater et Magistra* (Libreria Editrice Vaticana, 1961), p. 84: http://w2.vatican.va/content/john-xxiii/en/encyclicals/documents/hf_j-xxiii_enc_15051961_mater.html
55. Ibid., para. 148.
56. Pope John Paul II, "Promote Real Economic Democracy," address to the Central Institute of Cooperative Credit Banks (June 26, 1998): www.cin.org/jp2/jp980626.html

57. Pope Benedict XVI, "Address to Members of the Confederation of Italian Cooperatives and of the Italian Federation of Cooperative Credit Banks," Libreria Editrice Vaticana, December 10, 2011: https://is.gd/YGefwN
58. Of which I am a columnist.
59. Elinor Ostrom, "A General Framework for Analyzing Sustainability of Social-Ecological Systems," *Science*, 325 (2009), 419–22.
60. This is a challenge since most US electric co-ops are entangled in contracts that commit them to coal. Özgür Yildiz et al., "Renewable Energy Cooperatives as Gatekeepers or Facilitators? Recent Developments in Germany and a Multidisciplinary Research Agenda," *Energy Research and Social Science*, 6 (2015), 59–73; John Farrell et al., "Re-Member-ing the Electric," *Energy* (Mar. 29, 2016): https://ilsr.org/report-remembering-the-electric-cooperative/
61. Arizmendiarrieta, *Reflections*, § 180. For a contemporary example of technologically oriented cooperativism, see Trebor Scholz and Nathan Schneider (eds.), *Ours to Hack and to Own: The Rise of Platform Cooperativism, a New Vision for the Future of Work and a Fairer Internet* (New York, NY: OR Books, 2016).
62. See, published by the Democracy Collaborative, a leading advocate of the anchor institution model: "Anchor Institutions," *Community-Wealth.org*: http://community-wealth.org/strategies/panel/anchors/index.html
63. Pope Francis, *The Name of God Is Mercy* (New York, NY: Random House, 2016), p. 58.

9

Toward an Ethic of "Civic and Political Love" in the Workplace

Amanda Jaret and David L. Gregory[*]

Following the near collapse of the global financial system, transnational political and financial actors have begun to exercise ever greater control over local political and economic formations while continuing to demand rapid, sustained growth, with little regard for long-term ecological and social consequences. Structural unemployment and underemployment are increasingly pervasive. And, during the spring of 2015, the collapse of Syria set a devastating wave of mass migration in motion, with reverberations of genocidal reaction that could be felt throughout the Middle East and Eastern Europe. In this atmosphere, violence, exploitation, attacks on nature, and the abandonment of the most vulnerable are apparent everywhere. Undaunted in the midst of this scene of global chaos, Pope Francis issues *Laudato Si'*, manifesting a cautious, pragmatic view of the risks of ecological and social crisis and of contemporary labor.

Every encyclical highlights particular matters and special interests in furthering teaching authority and the social justice initiatives of the Catholic Church. So, too, *Laudato Si'*, situated at the heart of potentially global environmental disaster. Against this backdrop, the encyclical opens with a somewhat unexpected characterization of our relationship with the natural world. Francis describes the earth as not only our mother, but also our sister (LS 1–2). In a period of social, economic, political, and ecological crisis, this framing serves an important signaling function: it emphasizes the "relationship of mutual responsibility between human beings and nature" (LS 67, 92). By using this analogy, Francis also identifies a failure of fraternal love as a fundamental cause of our present climate and social crises. He argues that we have given ourselves over to a "tyrannical" and "distorted" anthropocentrism that encourages us to pay disproportionate attention to short-term economic gain without adequate regard for whether our activities are consonant with a proper conception of human dignity and the advancement of the common good and satisfaction of human needs (LS 68–69). By focusing on both our loss of fraternal connection with the natural world and with one another, Francis explicitly links the risks of ecological and social crisis. He argues that "[w]e are not faced with two separate crises, one environmental and the other social, but rather with one complex crisis which is both social and environmental" (LS 139).

Francis attributes our social and ecological crises to "rapidification," a term which he defines to include both an "intensified pace of life and work" and its accompanying web of accelerating technological, social, and economic changes (LS 18). Because rapidification contributes to both environmental deterioration and human degradation, Francis calls us to rethink the criteria that guide our economic and political decisionmaking. In place of a "deified market" that is dependent on relations of domination and an efficiency-driven, technocratic regulatory framework that only addresses the symptoms of our present crisis, Francis proposes "a change of humanity" to redress the effects of our twin ecological and social crises (LS 9, 18, 56). Specifically, Francis calls us to develop an "integral ecology" and urges us to forge a culture of care by engaging in distinctively fraternal acts of "civic and political love" (LS 92).

Because the encyclical is so closely attuned to questions of fraternity and mutuality, it is unsurprising that Francis highlights the centrality of human labor to the project of developing an integral ecology. As Francis recognizes, labor mediates our relationship with nature and enables humans to take part in the ongoing work of creation. And, labor is cooperative and puts us in contact and community with others. For these reasons, Francis identifies human work as both a vital aspect of caring for our common home and a way to develop meaningful, fraternal relationships with others and the natural world that will contribute to our flourishing.

This chapter explores the workplace as a site for developing an ethic of civic and political love that will lead towards an integral ecology. We begin by discussing the human vocation to work and the dignity of labor. We will then discuss the concept of rapidification and trace some of its manifestations in the workplace, including the fissuring and fracturing of the employment relationship and the tendency toward automation. Finally, we will explore how gestures of mutual care in the workplace that are consistent with the principles of solidarity and subsidiarity can help repair the damage caused by rapidification and offer a promising alternative to both the deified market and efficiency-driven, top-down decisionmaking.

1 "CREATED WITH A VOCATION TO WORK"

In *Laudato Si'*, Francis ably summarizes the core of more than a century of magisterial Catholic social teaching on the theology of work and the dignity of workers. Beginning with *Rerum Novarum*, Leo XIII's landmark 1891 encyclical, the Church has recognized that human labor is central to the "social question." As perhaps best expressed by John Paul II, "by means of work man shares in the work of creation."[1] Situating himself within a century of Catholic social teaching on work and following John Paul II's landmark encyclicals on labor, Francis affirms the dignity of labor and our vocation to work.

As we trace the path that Francis takes in the encyclical, we begin at the beginning. But then, even beginnings have beginnings. As Francis recounts, the Judeo-

Christian story of creation casts an important light on the dignity of labor. Francis explains that the biblical account of creation shows us that each person is created out of love and made in God's image and likeness. Because each human life is the product of a specific and deliberate, willed act of love, Francis argues that the story of creation demonstrates the infinite dignity of each individual. He also reminds us that we learn in Genesis that human beings were placed in the garden to preserve and "keep" it, but also to "till" the earth and render it fruitful (LS 65–66, LS 124).

However, as Francis explains, the biblical story of creation also teaches us that the rupture of original sin alters our relationship with the natural world, marking it as conflictual (LS 65–66). Francis also identifies the story of Cain's murder of his brother Abel as an illustration of the relationship between the failure of fraternal love and disharmony with nature. As Francis argues, "[d]isregard for the duty to cultivate and maintain a proper relationship with my neighbor, for whose care and custody I am responsible, ruins my relationship with my own self, with others, with God, and with the earth" (LS 70). Because of this disharmony, labor plays a particularly significant role in mediating the relationships among human beings and between humans and the natural world and can contribute to repairing the harm caused by sin.

Francis also offers a thoughtful account of how the story of creation is the penultimate chapter in the saga of sanctified work, which is only completed with the entry of Jesus, the carpenter of Nazareth, into public life. Francis recalls that Jesus was "far removed from philosophies which despised the body, matter and the things of the world," but instead that He "worked with his hands, in daily contact with the matter created by God, to which he gave form by his craftsmanship." In this manner, Francis explains, Jesus "sanctified human labor and endowed it with a special significance for our development" (LS 98, 124). Francis also notes that work is a special forum for establishing social ties and sustaining forms of human encounter that are less and less common in an isolated, atomized social context (LS 47, 49). As such, Francis situates labor as a vital aspect of an integral ecology and a way for human beings to find meaning in life on this earth, as well as a path to growth, human development, and personal fulfillment (LS 128).

II RAPIDIFICATION AND THE WORKPLACE

While Catholic social teaching has long recognized the centrality of work to the social question, conditions in the workplace have undergone transformative changes in the past few decades. From the 1940s until the early 1970s, workers in the United States often remained with one firm for the duration of their careers. In 1946, at the peak of union density, 37 percent of private sector employees were unionized.[2] And, after the so-called "Treaty of Detroit" between the United Auto Workers and General Motors, workers who selected union representation generally forfeited their right to strike for a period of several years in exchange for a collective-bargaining agreement that enshrined a panoply of employment benefits and on-the-job protections. Workers could expect to receive adequate wages, along with health and retirement benefits tied to their years of service with the firm, grievance-arbitration mechanisms, and provisions securing respect for seniority and protections against unjust termination.[3]

By the 1970s, international competition became more ferocious as the post-war period of American industrial dominance began to wane. Technological progress also contributed to a substantially increased mobility of capital, and employers in traditionally unionized industries began to relocate their productive operations to lower cost labor markets abroad. As union density began to decline, the historic arrangement set in motion by the Treaty of Detroit began to fray. In addition, as capital became more mobile, employees could seldom wield a sufficiently credible strike threat to meaningfully improve their bargaining position, especially after the employers increasingly began hiring permanent strike replacements during the 1980s.[4] These trends have continued unabated in the intervening years, and as of 2016, only 6.7 percent of private-sector employees in the United States are unionized.[5] As union density has declined, technological progress has increased output tremendously without a corresponding rise in job growth, and workers' standard of living has not kept pace with their contributions to heightened productivity.[6]

Beginning in the 1970s, and responding to the incipient shift in the character of working conditions in industrialized nations, Catholic social teaching began to emphasize the potentially injurious consequences of technological change for workers. In *Redemptor hominis* and *Laborem exercens*, Pope John Paul II criticized the manner in which technology encourages us to place productivity ahead of human dignity and can be used to replace human workers.[7] Francis sharpens this critique with his introduction of the term "rapidification." For Francis, rapidification is a form of acceleration that involves both straightforward technological advances and the less immediately apparent social changes in the tempo of life and work that technological change causes (LS 18).[8]

In discussing the effects of rapidification, Francis emphasizes the Janus-faced character of technological acceleration. Francis does not dispute that technology serves a valuable role in mediating the relationship between human beings and the natural world; in fact, he celebrates both the aesthetic value of technology and its practical ability to improve the quality of human life. However, he is also highly sensitive to the unevenness of the processes of rapidification, noting that when technological progress is not accompanied by social changes leading to "a development in human responsibility, values and conscience," such apparent progress may ultimately serve to reify existing relations of domination or cede control to individuals who understand how technology functions but may not be inclined to prioritize the common good. Francis provocatively contends that because of rapidification, "we stand naked and exposed in the face of our ever-increasing power, lacking the wherewithal to control it" (LS 102–103, 105).

Rapidification manifests itself in the workplace in countless forms. As an initial matter, rapidification is closely associated with the trend toward increasing financialization of the economy and theories of "data-driven" management. Increasing financialization is also associated with an acceleration of the processes of globalization and the opening up of markets through "free trade" agreements. All of these phenomena create hydraulic pressures that may compel even well-meaning employers to depress wages, slash benefits, and replace human workers with automated processes whenever possible.[9] Francis is quite adamant in his critique of the dynamics of financialization: he emphasizes that the financial system embodies "a power which has no future and will only give rise to new crises after a slow, costly and only apparent recovery" (LS 189). And, he observes, the financialization of the economy creates tendencies toward crises of overproduction and makes it less profitable to engage in productive, rather than merely speculative, economic activity (LS 109). These developments amount to a reversal of means and ends: by subjecting human persons to the principle of the maximization of profits, Francis argues that we "misunderstand the very concept of the economy," which is to satisfy human needs (LS 195).[10]

The technological and social acceleration that accompanies rapidification is strikingly apparent in the contemporary workplace. The accelerating application of automated and digitally mediated processes and the ascendancy of "just-in-time" and "lean" manufacturing models is the predictable consequence of the economic imperative to maximize short-term gain.[11] The speed of production in some industries has accelerated to the point of hazard. In one particularly gruesome example, workers at meatpacking plants report that they are routinely denied permission to take bathroom breaks because doing so interferes with the continuous operation of processing lines. Many workers have reacted by wearing diapers at work or by dangerously restricting their intake of food and water.[12]

As Francis explains, rapidification also entails an increasing velocity of social change. Perhaps one of the most distinctive recent developments illustrating the social effect of rapidification in the workplace is the process that David Weil and others describe as the "fissuring" of the employment relationship. The fissured workplace is characterized by a complex web of subcontractors and temporary employment agencies that operate in tandem to meet employers' immediate labor needs. The architecture of the fissured workplace is often intentionally designed to assist employers in obscuring or evading their legal obligations. So-called "user employers" who rely on labor supplied from subcontractors or employment agencies rely on numerous tactics to eschew responsibility for their leased workers. In another variation, employers may "fracture" the employment relationship by relying on temporary or contingent workers or misclassifying their workers as independent contractors to avoid legislation that protects statutory "employees" but exempts independent contractors from coverage.[13] In some instances, workers are ostensibly hired as temporary employees but may become "permatemps" who work alongside the user's ordinary employees for years at a time.[14]

The advent of the so-called "gig" or "on-demand" economy has further accelerated the fissuring and fracturing of the employment relationship. In the gig economy, companies operating software-based platforms promise to connect consumers with workers who can perform nearly any services. A certain level of technological development and pace of acceleration was the condition of possibility for the gig economy: firms that operate platforms for the purpose of mediating transactions for the performance of services are only able to connect workers with consumers because of the widespread availability of smartphones and the development of sophisticated software applications. Unfortunately, companies that operate such platforms often prefer to act as mere intermediaries, eschewing responsibility for the workers whose services are essential to the success of the platform.[15]

Perhaps the most worrisome long-term consequence of these acceleration phenomena is the rise of persistent unemployment and underemployment. Francis observes that the increasing emphasis on short-term economic gain makes it difficult for employers to resist shifting to automated production processes, which undermines the crucial goal of ensuring access to steady employment for everyone. And, he opines that the catastrophic unemployment of young people and the barriers to entry into the workforce are among the greatest challenges to social justice in the world today (LS 127–128). As employers increasingly automate functions that workers previously performed, workers with specialized skills that have become obsolete often find themselves structurally unemployed. Those who are unable to keep up with the increasing tempo of work or are displaced by automation may find it difficult to regain a foothold in the workplace.[16] Over time, unemployment and underemployment may effectively prevent people from fulfilling their vocation for work and exacerbate forms of alienation and indifference that are intimately bound up with the social crisis described in the encyclical.[17]

III CIVIC AND POLITICAL LOVE IN THE WORKPLACE

Francis conceives of rapidification as a symptom of our loss of regard for one another and for nature and an indication that relations of domination have begun to characterize relationships that should be founded on mutuality. In this regard, rapidification is one of the unforeseen consequences of the "misguided anthropocentrism" that Francis decries. Modern anthropocentrism not only creates incentives for us to consciously eschew responsibility for other persons and for nature, but it also sets in motion a "constant schizophrenia" and a culture of relativism that causes us to mistake other persons and the natural world for mere objects, using them as means rather than ends in themselves (LS 82, 115–119).

Instead of seeking domination and "dominion," Francis calls us to re-envision our relationship with nature and one another as one guided by "responsible stewardship" (LS 116). To combat rapidification and its accelerating tendencies, Francis urges us to "slow down and look at reality in a different way" and to return to the values and goals that have been swept away by our "unrestrained delusions of grandeur" (LS 114). Relatedly, Francis urges us to develop an "integral ecology" which is "inseparable from the notion of the common good." Following Pope Paul VI, Francis defines the "common good" as "the sum of those conditions of social life which allow social groups and their individual members relatively thorough and ready access to their own fulfillment" (LS 156).[18] For Francis, embracing the common good entails a rejection of anthropocentrism and a fundamental reorientation of our economic activity that will place human needs and development at the center.

What is perhaps most striking about the encyclical is the manner in which Francis recommends that we promote the development of an integral ecology that has the capacity to reorient our decisionmaking to privilege the common good. While we might expect a clarion call for transformative engagement directly with elites, Francis instead recommends that we begin with a more modest neighborhood dedication to addressing ecological and social problems at the local level. Specifically, he calls us to engage in acts of "civic and political love" that demonstrate care for nature and one another and contribute to our understanding of a "universal fraternity." He emphasizes that an "integral ecology is also made up of simple daily gestures which break with the logic of violence, exploitation and selfishness." Francis thus imagines that engaging in small "gestures of mutual care" and acts of fraternal love will impel us to "devise larger strategies to halt environmental degradation" because they will have spillover effects that "encourage a 'culture of care' which permeates all of society." And, as Francis reminds us, "[f]raternal love can only be gratuitous; it can never be a means of repaying others for what they have done or will do for us" (LS 228–231).

The workplace is a key site of contestation in the effort to establish an integral ecology, promote the common good, and contribute to a culture of civic and political love. This is because the workplace is a significant locus of vocational activity and a site for potentially transformative encounters with others and the natural world that can help us reestablish relationships of responsible stewardship and mutuality. Indeed, Francis explicitly argues that "[a]ny approach to an integral ecology, which by definition does not exclude human beings, must take account of the value of labor" (LS 124). Protecting the value of labor and the dignity of work is therefore a crucial component of reversing the ruinous tendencies of rapidification and anthropocentrism and ensuring that human persons are not treated as mere objects, but instead as the proper subjects of our economic and social ordering.

Taken this way, we see that Francis enjoins us to develop local capacity as a means of transforming global patterns of ecological and economic activity. This perspective is consistent with prior encyclicals and situated within the familiar framework of the Catholic social thought principles of subsidiarity, solidarity, and the preferential option for the poor. As such, we begin our discussion of the possibility of developing an ethic of civic and political love in the workplace by revisiting these principles and their application in the workplace before evaluating how Francis deploys these concepts in *Laudato Si'*.

As prior encyclicals teach us, subsidiarity is an ordering principle that advises that smaller and subordinate organizations should strive to fulfill human needs whenever possible and resists the accumulation of authority in larger and higher associations than are necessary.[19] The principle of subsidiarity is rooted in a conception of human dignity that recognizes that human persons are "ontologically and morally prior to the state or other social groupings."[20] For this reason, subsidiarity envisions persons as the primary agents of social action and the proper subjects of political and economic decisionmaking.

While subsidiarity maintains a focus on the centrality of individual persons as subjects who possess an inherent dignity, it does not counsel in favor of a wholesale devolution of political power to individual actors. Instead, subsidiarity acknowledges the value of "mediating institutions" that are capable of responding to human needs and addressing suffering. And in appropriate circumstances, the principles of subsidiarity suggest that such organizations may properly act as moral agents.[21] Past encyclicals teach us that certain intermediate organizations are particularly significant because they are proven incubators of the conditions that promote human flourishing (like the family or certain civic organizations). Other institutions, like labor unions and other worker associations, can act as moral agents when they are closely attuned to local problems and capable of forging solutions that are carefully tailored to resolving specific human needs. Subsidiarity also recognizes the role of state and local governments in establishing and maintaining the juridical framework that implements the natural right of freedom of association. And where merely local action is incapable of sufficiently coordinating action and responding to human needs, the principles of subsidiarity also justify more direct forms of state action and intervention in political and economic life.

By contrast, solidarity and the preferential option for the poor provide normative guidance for what our political and economic decisionmaking should prioritize. Solidarity is the "substantive vision of the human person in relationship with others."[22] As defined by Pope John Paul II, acting in accordance with the principle solidarity entails making a "commitment to the good of one's neighbor" and recognizing the dignity and moral agency of all persons.[23] As such, solidarity is a simultaneous commitment to the good of all and to seeking the highest good for each individual. This commitment requires more than "vague compassion or shallow distress";[24] it is a call to action that "binds us to make ourselves the neighbor of every person without exception, and of actively helping him when he comes across our path."[25] The preferential option for the poor, in turn, is a "special form of primacy in the exercise of Christian charity" that regards the most vulnerable as deserving of particularly solicitous moral consideration.[26]

As such, for Catholic social thought, solidarity and the preferential option for the poor help establish the appropriate ends of social policies, while subsidiarity provides us with the framework that is best suited to achieving those ends. It is worth stressing that these concepts are not merely prudential benchmarks for policymaking or equivalent with their secular counterparts. When Francis rejects the commodification of work and of workers in *Laudato Si'*, he does so in a manner that is distinctively Catholic because his discussion is rooted in a vision of the human person's dignity and the essential conditions that support human flourishing. Additionally, when he speaks of the importance of solidarity, he does so by identifying a natural human right to form voluntary associations.

Francis begins by discussing subsidiarity as a means of structuring our social institutions and then describes how our effort to attain the common good and to ensure the universal destination of the world's goods is "logically and inevitably" a summons to solidarity and the preferential option for the poor (LS 158). He specifically identifies the mediating function that social institutions, including the state, intermediate civil organizations, and the family can serve in the regulation of human relationships (LS 142, 157). Francis appreciates the channeling and aggregation functions that intermediate organizations can undertake, noting that "[s]ocial problems must be addressed by community networks and not simply by the sum of individual good deeds" (LS 219). Yet he also notes that uniform, top-down regulation is not always suitable to redressing local concerns and may become another form of domination and technocratic mastery. Instead, Francis predicts that an integral ecology is more likely to spring up with the application of dynamic, participatory processes that are geared towards securing the common good and advancing distributive justice (LS 144, 157, 196).

In addition, Francis explains the role of solidarity and the preferential option for the poor in our efforts to attain the common good (LS 158). His characterization of the call to solidarity is couched in unambiguously fraternal terms. In this connection, it is interesting to recall the analogy of "sister earth" that Francis uses at the very beginning of the encyclical. Francis lauds St. Francis of Assisi's practice of referring to all creatures as his brothers and sisters as an example of how acting in accordance with the conviction that we are interconnected and share a destiny will help defeat attitudes of mastery and exploitation (LS 11). Francis emphasizes that "genuine care for our own lives and our relationships with nature is inseparable from fraternity, justice and faithfulness to others" (LS 70).

If, following Francis, we envision solidarity as a form of fraternal love, the encyclical's call to action has particular salience in the workplace (LS 228). While Francis does not directly address the role of collective bargaining in the labor relations context, he advocates putting a well-trained eye on collective initiatives that reduce the harmful effects of rapidification and emphasizes the value of local participation in the inauguration of an integral ecology. He characterizes such initiatives in terms of the principle of the common good, arguing that "[t]o ensure economic freedom from which all can effectively benefit, restraints occasionally have to be imposed on those possessing greater resources and financial power" (LS 129).

Francis' sentiments regarding the role of solidarity in the workplace find a noticeable parallel in the principles that undergird American legal protections for workers' associational activities.[27] In the United States, the foundational statutory protection for private sector workers' freedom of association is the National Labor Relations Act ("NLRA" or "the Act"). Section 7 of the Act provides, in relevant part, that employees have the right "to form, join, or assist labor organizations, to bargain collectively through representatives of their own choosing, and to engage in other concerted activities for the purpose of collective bargaining or other mutual aid or protection[.]"[28] Section 1 of the Act indicates that it is the policy of the United States to "encourag[e] the practice and procedure of collective bargaining and . . . protect[] the exercise by workers of full freedom of association, self-organization, and designation of representatives of their own choosing[.]"[29] The statutory protection offered by the Act safeguards a wide range of associational conduct, including traditional organizing and bargaining activities and other forms of mutual aid and conduct that seek to promote solidarity. As the Supreme Court has explained, the NLRA's wide-ranging protection for varied forms of worker association and conduct evinces "a congressional intent to create an equality in bargaining power between the employee and the employer[.]"[30]

As we recall the principles of Catholic social thought, we will attempt to demonstrate how forms of worker associative activity which are consistent with the principles of subsidiarity and solidarity can contribute to the formation of an integral ecology. We offer some thoughts regarding some of the possible avenues for developing an ethic of mutual, fraternal care through worker associational activity. These examples are not exhaustive, but we hope they illustrate the potentially transformative possibilities that can result from local collective action.

First, the collective bargaining relationship may be an effective means of establishing substantive norms that are capable of counteracting the harmful effects of rapidification. Collective bargaining can be a particularly useful function for unions to perform as intermediate institutions within a framework of subsidiarity because it allows workers and their employers to come to mutually agreeable terms, with an eye toward unique local circumstances and needs. In this respect, unlike top-down, static forms of regulation, the form of collective bargaining that the NLRA sanctions is relatively dynamic and gives parties the freedom to determine what kinds of substantive contract terms they will adopt. Workers may pursue a wide range of contract proposals during bargaining, with an eye toward solving concerns that are specific to particular workplaces or workforces. For instance, workers might seek contractual protections against forms of discrimination that they have experienced in the past. One especially significant set of contractual terms workers might pursue includes employer leave policies. As Francis notes, rest gives us the time and space for contemplating the meaning of our work and dignifies our labor (LS 237). During bargaining, workers frequently seek inclusive family and medical leave policies and to ensure more generous pregnancy and maternity or paternity leave benefits than might otherwise be available. Additionally, collective-bargaining agreements typically contain grievance procedures that allow employers and workers to establish mutually agreeable mechanisms for resolving disputes expeditiously and at relatively low cost.

In addition, the workplace is also important because it is a site of productive activity. While workers are generally unable to bargain directly over entrepreneurial decisions, they may be able to influence the economic and ecological effects of industrial activity through the bargaining process. For instance, workers might bargain to establish more stringent health and safety standards that both protect workers and reduce the risk of industrial accidents that could also damage the natural world. Bargaining is not the only manner in which workers may seek to advance such goals: when workers have the freedom to organize and express themselves on the job, valuable opportunities to build alliances with other intermediate civil organizations that are seeking to protect the natural world may emerge. There have been a number of successful examples of this kind of organizing, perhaps best typified by the Oil, Chemical, and Atomic Workers' coordination of workers and community stakeholders who shared an interest in reducing the negative health and environmental effects of industrial pollutants.[31]

We could imagine other, more instrumental uses of the mechanism of collective bargaining that may contribute to a culture of fraternal love. For example, worker associations might use the structures of collective bargaining to resist two-tiered wage structures that fracture the workplace on generational lines. Francis specifically calls for intergenerational solidarity in the encyclical, and the workplace may be a fruitful site for building and maintaining intergenerational social cohesion (LS 159). All too often, however, the experience of rapidification in the workplace has affected younger and older workers differently, and the possibility of intergenerational conflict arises when workers who are part of the same bargaining unit discover that they have divergent interests. An ethic of civic and political love in the workplace might look like older workers resisting two-tiered wage structures and standing in solidarity with younger workers to ensure that the path toward full, remunerative employment they enjoyed remains available to future generations. Correspondingly, younger workers might help build intergenerational solidarity by defending older workers' promised retirement benefits, even if they will not enjoy similar benefits themselves. These forms of engagement are particularly exemplary models of the kind of gratuitous fraternal love Francis speaks about in *Laudato Si'* because they are so evidently other-directed.

While formalized, institutional worker associational activity comports with the principles of subsidiarity and can serve a valuable coordinating function, there are many opportunities for gestures of mutual aid in the nonunionized workplace to ripen into a culture of care and enduring forms of civic and political love. Courts have interpreted the NLRA to protect gratuitous forms of mutual aid, including sympathy strikes, concerted walkouts and protests, and other actions that are directed at "making common cause with a fellow workman over his separate grievance," even if "only one of them . . . has any immediate stake in the outcome."[32] Workers who have participated in actions like the fast food workers' "Fight for Fifteen" have noted the sense of dignity and respect they derive from giving and receiving mutual aid.[33] Acts of mutual aid and protection can serve as an effective means of combating the tendencies towards isolation and the rupture of the bonds of social cohesion that accompany rapidification (LS 46–47).

Worker associational activity is also closely aligned with participatory self-government, education, and the goals of industrial democracy.[34] Workers may take on leadership roles in unions and other worker associations, and their involvement in such activities may be helpful preparation for political participation or other forms of robust engagement with civil society. Francis calls on intermediate organizations to engage in pedagogical activities with an eye toward promoting integral ecology (LS 213–214). Unions and other labor organizations have often served an important role in educating their membership, and if they take the call to engage in ecological and economic education seriously, they may profoundly affect their members' views on production and consumption for the common good, which may have a transformational spillover effect on workers' families and communities.

Because participating in union activities or engaging in other concerted activity for the sake of mutual aid or protection is an expression of worker solidarity that seeks to advance the principle of the common good, worker organizations and associational activities are more than merely instrumentally valuable. Worker associations are closely related to the goal of safeguarding human dignity and are a special locus of associational activity and worker solidarity that can contribute to the project of subsidiarity and, in turn, help lead towards an integral ecology. By providing a forum in which workers can express their interests and exercise voice, collective bargaining helps us move closer to the kind of political and economic system that prioritizes human needs that Francis recommends we strive toward (LS 189). Specifically, equalizing bargaining power helps ensure that workers are not treated as mere objects, but instead as ends in themselves (LS 189).[35]

Some of the tactics associated with labor protest correspond well with forms of protest Francis suggests might be useful as we pursue an integral ecology. For example, Francis notes the potential benefits of using consumer boycotts to promote better economic and ecological practice, arguing that "[w]hen social pressure affects their earnings, businesses clearly have to find way to produce differently." As Francis calls on consumers to exercise a greater degree of responsibility, workers also have a pivotal role to play in educating the public and in protesting exploitative practices or policies that will adversely affect the natural world (LS 206).[36] The manner in which Francis conceives of worker solidarity and consumer actions as part of a struggle for the common good is significant because it views workers' conduct as a show of solidarity that is for the sake of others, not solely against their employers.

As such, as long as the use of such tactics is carefully tailored to the achievement of ends that promote the common good, they are consistent with the principles of Catholic social thought.

We also wish to anticipate some potential objections to the viability of worker associations as sites for the development of an ethic of civic and political love. First, some might argue that worker organizations have often stood in the way of social and ecological progress because they are inclined to protect traditional methods of production or to preserve work opportunities for their members at any cost. For this reason, they might conclude, unions and other similar organizations are unlikely to have the institutional commitment to an integral ecology that is essential to developing an ethic of civic and political love.

There are certainly incidents that validate the concerns regarding unions' stance on social and ecological matters. Unions have at times taken defensive positions that have exclusionary consequences for workers who are different from their members. And unions representing construction and building trades workers have at times spoken out in favor of projects that may have negative environmental or social consequences because such projects offer the promise of decent jobs in the short term.[37] Nevertheless, we believe that abandoning the potentially restorative function of worker solidarity actions and the organizations that support such actions because of improvident decisions particular actors have made is short-sighted. Unions and other worker organizations are consciously grounded in an ethic of fraternal love, and while they may fall short, they strive toward the goals of civic and political love.[38] And if Francis is right in predicting that small gestures of mutual care and fraternal love can transform our institutions and effect a change in humanity, we have reason to believe that worker associations can be transformed or redirected in the same manner when such transformation is called for.

Others might argue that while worker associations are potentially valuable mechanisms for promoting an integral ecology, our legal framework for protecting worker solidarity actions is inadequate. It is undeniable that more robust legal protection for workers' associational rights is necessary. Workers need to be assured that there are strong, expeditious remedies available when their employers interfere with their associational rights, and existing legal protections should be extended to ensure that more categories of workers enjoy full freedom of association. While there is still much work to be done, there have been successful efforts to clear space for workers to resist the effects of rapidification, as in the National Labor Relations Board's *Browning-Ferris Industries* decision, which seeks to provide workers in a fissured employment relationship with a legal framework that permits them to bargain with all entities who are "joint employers" who control their terms and conditions of employment.[39] Additionally, labor law scholars have only recently begun to pay closer attention to the natural, civil, and human rights lineage of workers' rights. This new conception of workers' rights has generated some interesting proposals for how to accord workers' rights legal status as human rights or to clothe particular forms of expressive activity with constitutional protection.[40]

IV CONCLUSION

The workplace is an appropriate place for us to begin moving towards an integral ecology because it is a forum in which fraternal encounters with others and the natural world are possible. As Francis acknowledges, expressing solidarity with other human persons and developing feelings of interconnectedness is a necessary prerequisite to concern for the rest of the natural world (LS 91–92). As we look ahead to the challenges that face our common home, we can take comfort in one of the most beautiful lines in the encyclical, in which Francis reminds us that "[f]or all our limitations, gestures of generosity, solidarity and care cannot but well up within us, since we were made for love" (LS 58). Francis is the quintessential heart on fire—the contemplative in action. We would do well to heed his call to "sing as we go," confident in our wish that "our struggles and our concern for this planet never take away the joy of our hope" (LS 244).

Notes

* The views expressed are those of the authors and do not represent the view of the National Labor Relations Board or the federal government. We are deeply grateful to Frank Pasquale, Lawrence Joseph, Jessica Rutter, Darren Glenn, Andrew Midgen, and Björn Westergard for improving our work by discussing the encyclical with us and offering insightful comments on an earlier draft.

1. Pope John Paul II, *Laborem Exercens* (1981), 25: http://w2.vatican.va/content/john-paul-ii/en/encyclicals/documents/hf_jp-ii_enc_14091981_laborem-exercens.html
2. Kate Bronfenbrenner et al., *Organizing to Win: New Research on Union Strategies* (Ithaca, NY: Cornell University Press, 1998), p. 2.
3. Nelson Lichtenstein, *The Most Dangerous Man in Detroit* (New York, NY: BasicBooks, 1995), pp. 279–81.
4. While the Supreme Court has permitted private employers permanently to replace strikers since 1938, permanent replacement was relatively uncommon until President Ronald Reagan replaced striking federal air traffic controllers in 1981: Joseph A. McCartin, *Collision Course: Ronald Reagan, the Air Traffic Controllers, and the Strike That Changed America* (New York, NY: Oxford University Press, 2011), pp. 343–45.
5. Because of the higher rate of unionization in the public sector, the overall rate of union membership in the United States is 11.1 percent (Bureau of Labor Statistics, "Union Members Summary," January 28, 2016: www.bls.gov/news.release/union2.nro.htm).
6. Erik Brynjolfsson and Andrew McAfee, "Jobs, Productivity, and the Great Decoupling," *New York Times*, December 11, 2012.
7. Pope John Paul II, *Redemptor hominis*, 15–16 (1979); Pope John Paul II, *Laborem exercens*, 5.
8. By acknowledging the multifaceted impact of technical acceleration, Francis's discussion of rapidification contains echoes of Hartmut Rosa's valuable insight that while technological progress is a necessary material basis for a general sense of social acceleration, it is also typically accompanied by other forms of rapid social change and an increased tempo of life. All of these forms of acceleration may take place at different rates, giving rise to uncomfortable phenomena of desynchronization: Hartmut Rosa, *Social Acceleration: A New Theory of Modernity*, trans. Jonathan Trejo-Mathys (New York, NY: Columbia University Press, 2015), p. 86.

9. Frank Pasquale, *The Black Box Society* (Cambridge, MA: Harvard University Press, 2015), pp. 35, 81.
10. In a time of austerity and retrenchment, previously public functions are increasingly privatized, subjecting an ever greater number of workers to the vicissitudes of market dynamics and consigning them to the consequences of rapidification: Monica Prasad, *The Politics of Free Markets: The Rise of Neoliberal Economic Policies in Britain, France, Germany, and the United States* (Chicago, IL: University of Chicago Press, 2006), pp. 55–60.
11. Rosa, *Social Acceleration*, p. 73.
12. Oxfam America, "No Relief: Denial of Bathroom Breaks in the Poultry Industry," May 9, 2016: www.oxfamamerica.org/static/media/files/No_Relief_Embargo.pdf
13. David Weil, *The Fissured Workplace: Why Work Became So Bad for So Many and What Can Be Done to Improve It* (Cambridge, MA: Harvard University Press, 2014), p. 10. The phenomena of the fissured and fractured workplace are also concomitant with a massive decoupling of jobs and benefits. By 2016, less than half of the private workforce participated in employer-sponsored retirement or medical benefit plans (Bureau of Labor Statistics, "Employee Benefits in the United States," July 22, 2016: www.bls.gov/news.release/ebs2.nro.htm).
14. Katherine V. W. Stone, *From Widgets to Digits: Employment Regulation for the Changing Workplace* (New York, NY: Cambridge University Press, 2004), pp. 68–69.
15. Valerio de Stefano, "The Rise of the Just-in-Time Workforce: On-Demand Work, Crowdwork, and Labor Protection in the 'Gig Economy,'" *Comparative Labor Law and Policy Journal*, 37 (2016), 471, 473–80.
16. Rosa, *Social Acceleration*, p. 86.
17. Vicki Schultz, "Life's Work," Columbia Law Review, 100 (2000), 1881, 1889 (noting that the loss of work can render people "politically and socially inactive as they have lost a sense that they can be efficacious in the world.").
18. Pope Paul VI, *Gaudium et spes* (1965), p. 26: www.vatican.va/archive/hist_councils/ii_vatican_council/documents/vat-ii_const_19651207_gaudium-et-spes_en.html
19. Thomas C. Kohler, "Labor Law: 'Making Life More Human' – Work and the Social Question" in Michael A. Scaperlandaand Teresa Stanton Collett (eds.), *Recovering Self-Evident Truths: Catholic Perspectives on American Law* (Washington, DC: Catholic University of America Press, 2007), p. 169 n.10.
20. Paolo G. Carozza, "Subsidiarity as a Structural Principle of International Human Rights Law," *American Journal of International Law*, 97 (2003), 38, 42.
21. Robert K. Vischer, "Solidarity, Subsidiarity, and the Consumerist Impetus of American Law" in *Recovering Self-Evident Truths: Catholic Perspectives on American Law*, p. 87.
22. Ibid.
23. Pope John Paul II, *Sollicitudo rei socialis* (1987), pp. 38–39.
24. Ibid. p. 38.
25. Pope Paul VI, *Gaudium et spes*, p. 27.
26. *Sollicitudo rei socialis*, p. 42.
27. While our analysis focuses on private-sector employees' rights, we note that most states and the federal government offer similar protections to their workers. In addition, public-sector workers enjoy certain additional protections that derive from the Constitution. A comprehensive discussion of public-sector employment is beyond the scope of this chapter.
28. 29 U.S.C. § 157. The statute also protects employees' right to refrain from engaging in any of the enumerated activities.
29. 29 U.S.C. § 151.
30. *NLRB* v. *City Disposal Systems, Inc.*, 465 U.S. 822, 835 (1984).
31. Steven Greenhouse, "Anthony Mazzocchi, 76, Dies; Union Officer and Party Father," *New York Times*, October 9, 2002.
32. *NLRB* v. *Peter Cailler Kohler Swiss Chocolates Co.*, 130 F.2d 503, 505 (2d Cir. 1942). This sentiment is also captured in the labor slogan that "an injury to one is an injury to all," which dates back to at the nineteenth century: James B. Atleson, *Values and Assumptions in American Labor Law* (Amherst, MA: University of Massachusetts Press, 1983), p. 206.
33. William Finnegan, "Dignity: Fast Food Workers and a New Form of Labor Activism," *New Yorker*, September 15, 2014.

34. The Labor Management Reporting and Disclosure Act of 1959 seeks to ensure internal union democracy by providing members with a statutory "bill of rights" that protects workers' speech and associational interests: 29 U.S.C. § 411.
35. Pope John Paul II takes a similar view of the potential for worker solidarity through labor unions to have a transformative effect on political and economic structures and to react against "the degradation of man as the subject of work": *Laborem exercens*, para. 8.
36. The so-called "Teamsters and Turtles" protests that united worker organizations with environmental groups during the demonstrations against the World Trade Organization Ministerial Conference of 1999 in Seattle provide a colorful example of this ethic. See Marc Cooper, "Teamsters and Turtles: They're Together at Last," *Los Angeles Times*, December 2, 1999.
37. For a recent example, see the AFL-CIO's statement on the Dakota Access Pipeline project. "Dakota Access Pipeline Provides High-Quality Jobs," September 15, 2016: www.aflcio.org/Press-Room/Press-Releases/Dakota-Access-Pipeline-Provides-High-Quality-Jobs
38. It is worth reflecting on the way in which worker associations' use of language reflects their aspiration toward fraternal love. It is no accident that union members refer to one another as "brother" and "sister" or that representatives who take on responsibility for other union members in their shops are called "stewards."
39. *Browning-Ferris Industries of California, Inc.*, 362 NLRB No. 186 (2015).
40. See, e.g., James A. Gross, "A Long Overdue Beginning: The Promotion and Protection of Workers' Rights as Human Rights" in James A. Gross (ed.), *Workers' Rights as Human Rights*(Ithaca, NY: Cornell University Press, 2003), pp. 1–6; James Gray Pope, "How American Workers Lost the Right to Strike, and Other Tales," *Michigan Law Review*, 103 (2004), 518, 521–22.

10

Laudato Si' and Augmented Reality: In Search of an "Integral Ecology" for the Digital Age

*Alessandro Spina**

1 INTRODUCTION

The latest Encyclical letter *Laudato Si', on care of the common home*[1] of Pope Francis, has been described by many commentators as a "green" document – perhaps this is owed to the fact that, for the first time in the history of the Catholic Church social doctrine, the central theme of an encyclical letter is the discussion of environmental concerns. But what is exactly the environment? One could think the environment consists of the natural elements and certainly the problems of the natural world such as loss of biodiversity, marine pollution and climate change feature extensively in the encyclical letter. However, a more complex and deeper understanding of the term "environment" seems to emerge from the 246 paragraphs of the letter. It would seem reductive and oversimplified to consider the environment as encompassing only the concrete physical and biological elements of creation. In reality, the term "environment" is characterized as the relationships that human beings establish with the external world, including with other human beings (LS 139). Social and cultural aspects of human life are part of the "environment"; they are the factors that shape our interactions with the natural world. As clearly described by Pope Francis: "The human environment and the natural environment deteriorate together; we cannot adequately combat environmental degradation unless we attend to causes related to human and social degradation. [...] Today, we have to realize that a true ecological approach always becomes a social approach [...]" (LS 48–49).

In this context, it assumes a special significance the focus that the encyclical letter reserves to the role of technologies.[2] The fact is that technologies embody the kind of relationships we want to have not only with nature but also with other human beings. Technologies, in fact, create the framework of our lifestyles and decisions, and construct the social fabric in which we are interwoven and the relationship we have with our external surrounding.[3]

These considerations are applicable to all sorts of technological innovations, from cars to chemical products, which, together with the increased power in the control of the natural environment, bring about also a modification in the social dimension. But the transformations of the "human environment" appear to be particularly evident with regard to the emerging field of digital technologies. These technologies rely on the recording in binary code of combinations of the digits 0 and 1 to represent words and images. Digital technologies enable immense amounts of information to be compressed on small storage devices that can be easily preserved and transported. With digital technologies, human communication capabilities, including the capability to process, retrieve and send information, are enhanced. When the environment is understood as a natural as well as social space, the digital "world" is not any longer a parallel dimension juxtaposed to the real world which we inhabit, but rather it overlays and is itself part of it.[4]

Encyclical letters are not works of economics, law or political science, and it would be a grossly negligent mistake to consider recommendations contained in these documents as directly applicable to concrete policy issues[5] or to seek coherent and complete explanation of current affairs. Yet, the *Laudato Si'* seems to provide conceptual keys to understand the most pressing challenges that our society faces with regard to technologies and in particular digital technologies. Covering a wide array of environmental issues such as climate change, waste, the loss of biodiversity, genetic manipulation, the encyclical letter, in its vast intellectual reach, identifies two general concepts: the "technocratic paradigm," which corresponds to a self-referential and egoistic logic of dominium of the external world, and the "integral ecology," which aims to include ethical concerns in the relationship with the environment. They represent two different approaches to science and technology.

The aim of this chapter is to explore the aspects of the *Laudato Si'* that relate to the way technologies mediate man's relationship with the external world, in particular by focusing on a specific type of digital technology called Augmented Reality (AR). This chapter submits that there is an urgent case for giving effect to Pope Francis's call for an ethical assessment of the development and use of digital technologies that is advanced in the "integral ecology" model. Finally, the chapter argues that while the regulatory framework of technologies has traditionally focused on the natural externalities connected to the use of the technologies, the 'integral ecology' concept urges to integrate *ex ante* considerations about the possible behavioural consequences of technologies in order to shape their development.

The chapter is composed of three parts. After the Introduction, the idea that the environment in the *Laudato Si'* also encompasses social and cultural aspects of the external environment inhabited by mankind is analysed in more detail. The first part also describes the concept of integral ecology that underlines the message of Pope Francis regarding the role of technologies in the relationships between man and nature. The second part looks at how the integral ecology can be declined in the field of digital technologies and identifies some existing ideas that from a legal point of view seem to legitimize the public contestability of technological development. In the third part, one strand of digital innovation, i.e. AR is examined by taking as a case in point the mass video game Pokémon Go, which blends together for players online and offline spaces. The Conclusion discusses how the *Laudato Si'* could help in framing public debates about the ethical development of new digital technologies and raising awareness for their environmental impact.

II "EVERYTHING IS INTERCONNECTED": GOING BEYOND THE NATURAL DIMENSION OF THE ENVIRONMENT IN THE *LAUDATO SI'*

The social dimension of the environmental crisis is one aspect that the encyclical letter brings to the attention of the readers often times. The Pope reminds us that "Everything is interconnected." This is not only for the reasons that have caused the environmental crisis we are facing (LS 70, 117), but it is also rooted in the idea that forward-looking solutions to many environmental issues rest on a new holistic approach called "integral ecology" (LS 137, 240). This model aims at presenting the interconnectedness between all human activities and between human and non-human forms of life. In this constellation, the environment is not something standing apart of man, an external element to be dominated. It is the space that gives existential significance to mankind not only physically but culturally. Man does not only live on earth, but it has a "relationship" with it. Indeed, integral ecology implies that problems affecting the conditions of individuals, for example poverty or profound economic inequality will also have an impact on the environment and vice versa, that the environmental problems have an impact on social issues – for example, the destruction of natural habitats is also one the causes of impoverishment of the habitat of developing countries helping spiralling the mass migration from poor to richer areas of the world. The idea of "integral ecology" means that environmental issues should be addressed in a holistic perspective taking into account the potential impact to all persons concerned.[6] It follows from this that all human activities are potentially affecting the environment in that they modify the existing reality (from producing a social report to the design of a new product), and that as different aspects (physical, chemical, biological) of the planet are interrelated and all living species are part of what one could call a cognitive, "network", "fragmentation of knowledge and the isolation of bits of information can actually become a form of ignorance, unless they are integrated into a broader vision of reality" (LS 138). A validation of this link between natural and social aspects of the environment can be seen, for example, in the study made by Amartya Sen on famine and starvation.[7] He investigated the social context and factors of famines, through the concept of legal "entitlements." He found that there has never been a famine, as opposed to food shortages, in social systems with democratic elections and free press. This is because in a social context where people can hold government to account, the redistribution of food would be organized in a way to allocate food so that no one starves. It follows that the social consequences of a natural phenomenon are not predetermined and immutable but depend on the social infrastructures that regulate the relationships of the affected population.

Within the framework of an "integral ecology" model, the role of technologies is essential as they mediate our relationship with the human and natural environment and they embed ethical concerns and values.

The Pope does not wish to cast a negative light on the role that technologies have in our modern society. On the contrary, in numerous parts of the encyclical letter, the Pope holds a very positive belief in scientific progress and technological innovation. There are words of admiration in the encyclical for the products of technological development: "steam engines, railways, the telegraph, electricity, automobiles, aeroplanes, chemical industries, modern medicines, information technology and more recently, the digital revolution, robotics, biotechnologies and nanotechnologies. [...] Technology has remedied countless evils which used to harm and limit human beings. How can we not feel gratitude and appreciation for this progress, especially in the fields of medicine, engineering and communications?" (LS 102) or for the scientific profession whose open-ended mission to push always further the boundaries of knowledge is almost linked to the freedom of artists.[8] Moreover, it is acknowledged that "green" technologies can be very helpful in addressing environmental problems.[9]

We will see in more details how this "integral ecology," which introduces a much wider perspective, is functional for assessing the overall environmental impact of digital technologies. It suffices to say at this stage that in accordance with the integral ecology approach the environmental consequences of digital technologies would not only be measured in terms of physical factors such as the pollution from manufacturing or disposing hardware components[10] or from the energy consumption for powering computers but also in terms of the changes in the social practices made possible by software applications, for example what kind of human behaviour is facilitated by certain features of social networks towards the natural world or other human beings.

Perhaps one of the most perceptive and original reflections of the encyclical letter is the observation that "technological products are not neutral" for they embed and propagate political and social values.[11]

This is not something entirely new in science and technology studies,[12] and it will be shown how this observation has inspired much of the current legal scholarship on regulation of technology. Yet it is nonetheless surprising to see such a theory accepted for the first time in an encyclical letter. It is by relying on this observation that the Pope argues that it remains essential to direct technology development at the service of another type of progress, one that promotes a healthier and more sustainable lifestyle. In this context, careful urban planning and policies to favour public transport can be very effective instruments for the goal to reduce the level of pollution and the consumption of non-renewable energy (LS 153). The non-neutrality of technology development can be seen, for example, in the self-interested choice, made by manufacturers, not to invest in extending the lifespan of marketed technologies hence to force users to buy new products continuously.

As technologies are not neutral, their functionalities often embed choices which have a direct impact on the human and natural environment. In the encyclical letter a warning is made against the "technocratic paradigm." This is the form of unaccountable and uncontrolled power of technologies to dominate the environment as a mere instrumental good, a resource that man can exploit without boundaries: " it is the way that humanity has taken up technology and its development according to an undifferentiated and one-dimensional paradigm. This paradigm exalts the concept of a subject who, using logical and rational procedures, progressively approaches and gains control over an external object. [. . .] It is as if the subject were to find itself in the presence of something formless, completely open to manipulation." (LS 106). The technocratic paradigm maintains an ideological "faith" in the myth of progress and infinite economic growth. According to the "ironclad logic" of the technocratic paradigm, economic growth and technological development will be able to solve each environmental problem. In this logic, every environmental problem can be fixed with technical solutions, regardless of whether they could adversely affect other parts of the human or natural environment. This can be seen for example in some initiatives purporting to produce urgent and partial responses to environmental problems,[13] which often simplistically shift, the environmental or social burden to others.

III THE CONTESTATION OF THE DEVELOPMENT AND DESIGN OF DIGITAL TECHNOLOGIES IN LEGAL SCHOLARSHIP AND IN THE *LAUDATO SI'*

Digital technologies make a rapid appearance in the encyclical letter (at LS 47 only). They are mentioned to highlight the risk that people may use them as a shield against direct social contacts and the complexity of personal experiences. The Pope warns that their excessive use could bring about "a sort of mental pollution."[14] In this perspective, it is illuminating to observe that, in the digital world, social media are powerful tools to shape the cultural settings and affect the way in which we communicate with each other. But, more importantly, digital technologies are gradually integrated in every activity of our life and also in the physical spaces where we live. Aptly, the term "*on*life"[15] has been coined to describe an existence that relies on an immersive ecosystem of integrated and always-on, hyperconnected digital applications such as smartphones, health and fitness apps, social media, and search engines. The pollution represented by digital technologies is not only physical as represented by e-waste, toxic hardware relicts and silicon mines but a moral and social one.[16] It is reflected in the loss of the private sphere where we can explore diversity and seek a mental space for self-determination, in the reduction of diversity of perspectives and in the standardization of our emotions which are quantified, monitored, and eventually manipulated for commercial purposes.[17]

There is, therefore, a more profound lesson which can be drawn from the encyclical letter and that could affect our thinking about the relationship between digital technologies and the environment. The fact that they embed cultural choices and values means that they can influence substantially the behaviour of users not only in their interpersonal relationships but also vis-à-vis the natural world. The availability of information to which we are exposed through the Internet and social media shapes both our believes and behaviour so as to trigger us to act by concentrating our attention to specific facts or ideas In fact, the whole digital economy is premised on the creation of an ecosystem where information is abundant and human attention is scarce. A visible trend of the potential negative consequences of this form of digital pollution can be seen in the phenomenon of "fake news," a type of hoax or propaganda used to mislead public opinion for hidden political or commercial gains. The problems connected to the effect on human agency of the quality of the digital infosphere will certainly be further boosted by applications of AR. This technology superimposes a computer-generated image or information on a user's view of the real world, therefore merging online and offline dimensions of our existence.

The discussion above raises points that are not entirely new in the legal scholarship on digital technologies. The application of an integral ecology approach as expounded by the Pope in the encyclical letter *Laudato Si'* is consonant to various theoretical frameworks emerging from legal doctrine.

The first area is that of "cultural environmentalism,"[18] an idea that described the similarity between the rise of the concept of environmentalism and that of digital activism. While the former started in the 1950s as a consequence of the awareness that the then rights on property were not adequate to control environmental externalities, the latter emerged as a result of the Internet to traditional social and legal rules for governing cultural production, in particular with regard to intellectual property rights and affiliated rights. The basic tenet of this theory relates to the protection of the spaces of cultural experimentation and consumption online.

The second strand can be traced to the idea that "code is law," which was originally developed by Lawrence Lessig in his famous book of 1999.[19] The thesis exposed in the book argued that the technical architecture of which software code is made is a major source of regulation as it determines limitations on the actions that users can make. This idea that can be traced back to the concept of regulation by design which is not a novel finding[20] but in the digital era assumes a much more powerful connotation. In opposition to the restriction and forcing of behaviour in the physical reality, the way in which software code works is often invisible and unaccountable to passive users.

Finally, the third scholarship theme linked to exploring the effect that choices made in the development of digital technologies is the critique of the social consequences of Big Data and "algorithmic regulation."[21] The capacity to analyse and extract insights from enormous amount of data that we generate in our daily life has opened new possibilities for regulating and predicting human behaviour. The over-reliance on mathematical models to analyse and regulate human and social dimensions opens the door to the risk of bias and unfairness being disguised as incontestable outcome of alleged data-driven "scientific" analysis.[22]

What the legal scholarship and the encyclical letter point to is to reclaim a legitimate space for democratic or at least public *contestation* about digital technologies. Digital technologies create social externalities that should be adequately internalized by society through public oversight not only of their use but of their *design and development*. This process aimed at assessing *ex ante* the environmental impact is generally accepted in the case of technologies or industrial projects which affect the natural environment. By making the environment a much wider concept than that of natural space, the encyclical letter seems to provide a solid justification for extending the process of an *ex ante* public accountability of digital technologies. It follows that, in addressing the issue of the social consequences of digital technologies, we cannot provide a value judgment about their alleged intrinsic worth without having fully assessed what kind of social and human impact they could have. In essence, if we accept an idea of environment that goes beyond the realm of the physical space as it appears to be suggested by Pope Francis, we need to think about regulating the risks arising from the social externalities of digital technologies. Depending on their design and functional features, digital technologies could be instrumental in furthering the objective of the "integral ecology" model, for example in those cases in which they help harnessing the dispersed knowledge of communities and enable the communication and participation of individual. However, they are also being used for practices of exploitation of human sentiments for the purpose of commercial marketing or for standardizing human relationships with other human beings and the environment. The case of AR which will be discussed below is paradigmatic because it shows in a very tangible way the capacity of digital technologies to transform our natural environment by overlaying on it an artificial and digitally controlled vision of the reality. As for physical human artifacts made of concrete, the set-up of software code is able to interfere with the environment.

It is of a great concern that the standard methodology adopted by engineers and software developers to design and create online environments[23] is almost exclusively geared to attract users' attention and assuage their emotional decision-making process. This methodology for development of digital technologies is not subject to public accountability or independent scrutiny.[24] For example, the standard practice for software development is linked to concerns such as usability/navigability etc. with no assessment of whether it results in the promotion of certain users' behaviours. The paradigm of an integral ecology of the *Laudato Si'* would provide a robust intellectual and moral justification for a more proactive control of the features shaping human behaviour which are embedded in digital technologies.

IV POKÉMON GO AND THE NEW FRONTIER OF DIGITAL ENVIRONMENTALISM

So far in this chapter, it has been noted that digital technologies, including social media and communication technologies, are having a profound impact not only on the human and social dimension of our lives, but also on the environment. There is, however, a digital application which could be singled out for the concrete way in which it demonstrates how digital technologies modify the human relationship with the natural environment. AR is the superimposition of different informational levels to the experience of reality. The human perception of the external environment comes to be modified and "augmented" by the integration in our cognitive processes of connected devices that overlay in real time the visual experience with digitally curated information. Although speculations about AR and other immersive applications have been around for some time, it is the release in July 2016 of Pokémon Go, a video game based on AR, which has commodified this technology to the wider public. In this videogame to be used on smartphones, players aim at catching, battling and training virtual creatures (the pokémon), which are superimposed on the players' environment through the use of GPS technology and device cameras. The commercial success of this AR video game has been altogether astonishing: within few weeks from its release, an impressive 100 million users worldwide have played the video game. This has triggered a global frenzy with some notable cases of compulsive behaviour to play widely reported in the press: some players have suffered injuries,[25] or even death[26] in order to capture the virtual creatures in the physical world, a number of trespasses in private spaces[27] and inappropriate behaviour in public spaces. One could dismiss Pokémon Go as a trivial application of AR, but its huge success is perhaps evidence of the pervasive, and almost addictive, nature of AR as a new human experience which blurs the boundaries between the physical and the digitally curated environment. Other examples of AR applications still under development are the wearable connected glasses developed by Google, or the possibility to implant an AR device directly on to the retina of the eye, making it invisible to people other than the user.

The essential structure of AR rests on a bi-directional information flow.[28] There is an input information flow in the form of data that the device is taking in – for example, in the case of Pokémon Go, the users allow the app to collect geo-spatial information and the images taken by the camera. There is an output information stream in the form of information that the device is programmed to superimpose to the reality: in the case of Pokémon Go, the natural world is augmented by the pokémon, the virtual creatures that players need to capture.

There are a number of issues of a legal nature that arise from the use of Pokémon Go or other AR mechanisms. They have been broadly placed in two categories, safety and privacy. With regard to safety, it has been noted that players are exposed to dangers arising from the very nature of the game and it is discussed whether the game developer could be held liable also for distracted, excessive, irresponsible use of the game. The current terms of service of Pokémon Go include a suitable disclaimer of liability, but what about extreme location? There appears also an emerging liability connected to the choice of the location in which pokémon are placed: it is true that the choice of the location of pokémon is programmed by an algorithm that sets the on the basis of various factors, but the game developer could still be held responsible for setting out the specifications for the algorithm to operate. This could in itself raise the responsibility of the game developer vis-à-vis potential discrimination for the distribution of pokémon in the external world.[29] Moreover, it is questionable whether there is a right for the owner of a physical space to exclude third parties to use it in order to overlay the digital information and whether using the physical space to integrate virtual information would amount to trespass under common law. With regard to privacy, several commentators have expressed concerns as to the enormous amount of data collected by the app including geo-spatial information but also lifestyle information. With advanced analytics, the amount of data collected by the app during the game could give the app developer a very detailed profile of users for which they might or might not be aware. The concerns stemming from the use and diffusion of AR devices are not limited to the legal sphere. The rapid spread of AR, as epitomized by Pokémon Go, raises broader and deeper public policy considerations. The more fundamental concerns from a public policy perspective relate to the design and configuration of the game. In this context, it is fair to claim that there is a legitimate need to explore the environmental impact of the release of AR products. This is necessary to address in advance, at the time of development, the behavioural consequences of the use of AR technology. The holistic approach of integral ecology expounded by Pope Francis in the *Laudato Si'* could offer a solid intellectual ground to develop further this claim.

In order to make a full assessment of the benefits and risks of Pokémon Go, we would need to look at the practical behavioural consequences of the use of the game. Signs that software developers of Pokémon Go had considered partially this aspect can be seen for example by observing that the application stops working if the player is too fast, this is in order to prevent players from using the game while driving. The developers had considered a possible negative behavioural consequence of the AR technology, and had implemented some technical features to avoid it. Moreover, researchers at the Johns Hopkins Bloomberg School of Public Health have looked at the (unintended) benefits of the game. They describe the game as a useful "health app" with potential to prevent non-communicable diseases such as obesity and promote physical activity[30] engaging people to spend time outdoors and socializing with a potential beneficial unintended consequence in reducing stress anxiety and depression.[31]

The public debate accompanying the release of Pokémon Go helps in framing the analysis of an "integral ecology" when applied to AR technologies. On one hand, there is an aspect of individualism in the ambition to reduce the external reality to a set of information that is predefined and made salient only for a certain purpose. Some applications of AR, if designed in such a way as to present to the user discriminatory or illegitimate information, would reduce rather than augment the informational context in which we make decisions. They might manipulate the perception of the external reality or of other human beings, leading to forms of social blindness and encouraging the dynamic of creating a self-referential artificial space or "bubble" by reducing any contact with persons or opinions we do not like.[32] This would take the use and development of AR in the framework of that "technocratic paradigm" highlighted by Pope Francis- and appears as another tool to dominate and expand the power of those that have the knowledge and the economic resources to use this technology over nature and humanity. AR would then become another step for an anthropocentrism that compromises the intrinsic dignity of the world (LS 115) and reduces nature "to an object of utility, as raw material to be hammered into a useful shape."[33]

But an approach based on the vision of an integral ecology could bring a different perspective on certain applications of AR that can help users to act in a way that is in accordance with the natural equilibrium and the common good and can bring positive benefits to humankind by encouraging actions and knowledge rather than mere passivity. AR can help users to discern the elements of reality that we are more interested to discover and learn, hence enlarging, not restricting, our knowledge of the mystery of creation through visualization.[34] The relatively harmless purpose and effect of the AR game should not make us forget that the real issue at stake here is not a question of whether Pokémon Go is good or bad in ethical or utilitarian terms but whether there is a legitimate public duty to assess and control the social externalities of AR technologies and the impact they have for the environment into which they are released. The reactions which have accompanied the popular success of a videogame based on AR show that only a principled, ethical paradigm would be able to reconcile the contradictions posed by the digital transformation of reality.

This should not lead to discourage or hamper digital innovation but to facilitate a development which serves the common good and takes into account its environmental effects. AR reminds us of the incredible power that digital technologies have in transforming the relationship between man and nature, by shaping the salience of the information visualized in the reality and by providing knowledge of the external world.[35]

V CONCLUSION

The thoughtful and in-depth considerations of the *Laudato Si'* cannot be fitted into a straitjacket moral or political systematization. The *Laudato Si'* remains a document containing a profound spiritual reflection. But its message also calls for a deeper understanding of the relationship we have with the environment, broadly interpreted to encompass not only the natural elements but also the social and human dimension. It does not advocate for any specific public policy measure. It does, however, invite its readers, including legal scholarship, to review certain understandings and dogmatic assumptions. Moreover, it challenges legal scholars to seek, with a creative mind and a much broader perspective, solutions to problems. The responsibility of mankind towards the environment is not an abstract concept. It is closely linked to material facts such as the education, the lifestyle and the cultural options we choose and reflected in the laws and regulations that as a society we adopt. Technologies are not neutral as they give substance, in a tangible form, to the kind of interactions we have with the external world. The legal framework which has accompanied the commercialization of the technologies and the applications of the Industrial Revolution seems not to be well equipped to address the challenges of digital technologies as that focused mainly with natural externalities. The social externalities brought about by digital technologies cannot be measured and controlled by regulatory systems fitted for the control of externalities affecting exclusively the natural environment. It is the message that "everything is interconnected" which should make us think about the idea that as the world which we inhabit is increasingly digital or digitally "augmented," the term "environment" should capture a more complex dimension than that of the natural space. In this context, the pervasiveness of digital technologies in our lives, and their profound influence at a cognitive level, invites us to consider very carefully the environmental consequences that they could have. In the case of AR, the physical world is overlaid with information, modifying and manipulating the cognitive perception of reality. This interference with the environment calls for a careful assessment of its positive and negative impacts. The integral ecology approach offered by the Pope is particularly helpful to reconsider the possibility to contest and keep into account the design and development of digital applications. AR technologies are bound to transform both the human and natural dimension of the environment consequences, the assessment of the environmental impact cannot be performed without taking into account the ethical aspects of the transformations in the real world. The physical space cannot be simply "colonized" by AR developers, as the physical and virtual world are interconnected and overlaying information on reality means transforming that "web of relationship" that is, in the final instance, the environment.

Notes

* Member of the Legal Service, European Commission and visiting lecturer in the Faculty of Law, University of Fribourg. The opinions expressed in this chapter are solely the author's and do not reflect those of any institution. I would like to thank Alberto Alemanno, Laura Ammannati and Frank Pasquale for their valuable comments and suggestions. This work is dedicated to my daughter

Matilde: "Let us sing as we go. May our struggles and our concern for this planet never take away the joy of our hope" (LS 243).

1. Pope Francis, *Laudato Si'*.
2. I have already attempted to describe the importance that science and technologies assume in the encyclical letter in another article: Alessandro Spina, " Reflections on Science, Technology and Risk Regulation in Pope Francis' Encyclical Letter *Laudato Si'*," *European Journal of Risk Regulation*, 6 (2015), 579–85.
3. "It can be said that many problems of today's world stem from the tendency, at times unconscious, to make the methods and aims of science and technology an epistemological paradigm which shapes the lives of individuals and the workings of society. The effects of imposing this model on reality as a whole, human and social, are seen in the deterioration of the environment, but this is just one sign of a reductionism which affects every aspect of human and social life. We have to accept that technological products are not neutral for they create a framework which ends up conditioning lifestyles and shaping social possibilities along the lines by the interests of certain powerful groups" (LS 107).
4. Antonio Spadaro, *Cybertheology: Thinking Christianity in the Internet Era* (New York, NY: Fordham University Press, 2014).
5. W. J. Wagner, "Unlocking Catholic Social Doctrine: Narrative as Key," *Journal of Social Catholic Thought*, 7 (2010), 289, 307.
6. "Today the analysis of environmental problems cannot be separated from the analysis of human, family, work-related and urban contexts, nor from how individuals relate to themselves, which leads in turn to how they relate to others and to the environment" (LS 141).
7. Amartya Sen, *Poverty and Famines: An Essay on Entitlement and Deprivation* (New York, NY: Oxford University Press, 1981).
8. "Human creativity cannot be suppressed. If an artist cannot be stopped from using his or her creativity, neither should those who possess particular gifts for the advancement of science and technology be prevented from using their God-given talents for the service of others" (LS 131).
9. With regard to climate change and the technologies that could ameliorate this problem: "Worldwide there is minimal access to clean and renewable energy. There is still a need to develop adequate storage technologies. Some countries have made considerable progress although it is far from constituting a significant proportion. Investments have also been made in means of production and transportation that consume less energy and require fewer raw materials, as well as in methods of construction and renovating buildings which improve their energy efficiency" (LS 26).
10. Eric Williams, "Environmental Effects of Information and Communications Technologies," *Nature*, 479 (2011), 354–58.
11. "We have to accept that technological products are not neutral, for they create a framework which ends up conditioning lifestyles and shaping social possibilities along the lines dictated by the interests of certain powerful groups. Decisions which may seem purely instrumental are in reality decisions about the kind of society we want to build" (LS 107).
12. This theory is firmly rooted into Science and Technology Studies, however it was considered provocative in the seminal scholarly contribution of 1980 written by Langdon Winner: Langdon Winner, "Do Artifacts have Politics?" *Daedalus*, 109 (1980), 121–36.
13. "To seek only a technical remedy to each environmental problem which comes up is to separate what is in reality interconnected and to mask the true and the deepest problems of the global system" (LS 111). In other parts of the encyclical it has been pointed out that in ever more polluted and dirty urban spaces, green and "ecological" areas are created whose access is however limited to privileged few (LS 45) or the example of the increasing use of air conditioning (LS 55).
14. "Furthermore, when media and the digital world become omnipresent, their influence can stop people from learning how to live wisely, to think deeply and to love generously. In this context, the great sages of the past run the risk of going unheard amid the noise and distractions of an information overload. Efforts need to be made to help these media become sources of new cultural progress for humanity and not a threat to our deepest riches. True wisdom, as the fruit of self-examination, dialogue and generous encounter between persons, is not acquired by a mere accumulation of data which eventually leads to overload and confusion, a sort of mental pollution" (LS 47). Similarly, the

encyclical letter puts great emphasis on the "consumerist" lifestyle, which is actively promoted by economic interests and generates, with its throwaway culture, vast amount of waste and pollution.

15. Luciano Floridi, "A Look into the Future Impact of ICT on Our Lives," *The Information Society*, 23 (2007), 59–64, 62 and further elaborated in Luciano Floridi, *The Fourth Revolution: How the Infosphere Is Reshaping Human Reality* (Oxford: Oxford University Press, 2014), pp. 1–272.
16. This observation echoes the idea of a "Grey Ecology": the by-products of the digital era are threatening the ecosystem of our human and mental existence in the same way as the by-products of the Industrial Revolution affected the natural environment: Stefana Broadbent and Claire Lobet-Maris, "Toward a Grey Ecology" in Luciano Floridi (ed.), *The Onlife Manifesto: The Online Initiative* (Springer, 2015), p. 120.
17. Reference is made to the practices of social media platforms to study through data proxies the emotional status of their users in order to assess strategies to influence their online or offline behaviour. The research in this field is often conducted secretly but in 2014 an experiment of emotional contagion was published by a team of researchers working with Facebook data: Adam D. I. Kramer, Jamie E. Guillory and Jeffrey T. Hancock, "Experimental Evidence of Massive-Scale Emotional Contagion through Social Networks," *Proceedings of the National Academy of Sciences*, 111 (2014), 8788–90.
18. James Boyle, "A Politics of Intellectual Property: Environmentalism for the Net?" *Duke Law Journal*, 47 (1997), 87–116.
19. Lawrence Lessig, *Code and Other Laws of Cyberspace* (New York, NY: Basic Books, 1999), 1–297.
20. Karen Yeung, "Towards an Understanding of Regulation by Design" in Roger Brownsword and Karen Yeung (eds.), *Regulating Technologies: Legal Futures, Regulatory Frames and Technological Fixes* (Oxford: Hart, 2008), pp. 79–107.
21. Frank Pasquale, *The Black Box Society. The Secret Algorithms that Control Money and Information* (Cambridge, MA: Harvard University Press, 2015).
22. Zeynep Tufekci, "Algorithmic Harms Beyond Facebook and Google: Emergent Challenges of Computational Agency," *Colorado Technology Law Journal*, 13 (2015), 203–18; Cathy O'Neil, *Weapons of Math Destruction* (New York, NY: Crown, 2016), pp. 1–260.
23. Mireille Hillebrandt, "Law as Information in the Era of Data-Driven Agency," *The Modern Law Review*, 79 (2016), 6.
24. A notable exception of this general omission of a public scrutiny in the development of online applications can be seen in the new EU data protection legislation: Regulation (EU) 679/2016 (also widely known as the General Data Protection Regulation or by its acronym GDPR). This new piece of legislation, which entered into force in May 2018, contains an obligation, laid down at Article 25, for "controllers" of personal data to adopt "privacy-by-design" and "privacy-by-default" measures at an early stage of configuring new data processing operations. This means that future products and services take privacy requirements into account at the time of their development.
25. Hayley Tsukayama, "Pokemon Go's Unexpected Side Effect: Injuries," *Washington Post*, July 10, 2016: www.washingtonpost.com/news/the-switch/wp/2016/07/08/pokemon-gos-unexpected-side-effect-injuries/
26. Roland Shichijo, "Japan Records 1st Pokemon Go-related Death," *Tokyo Reporter*, August 24, 2016: www.tokyoreporter.com/2016/08/24/japan-records-1st-pokemon-go-related-death/
27. Cara McGoogan, "Pokemon Go Makers Sued for Encouraging Trespass," *Telegraph*, August 3, 2016: www.telegraph.co.uk/technology/2016/08/03/pokemon-go-makers-sued-for-encouraging-trespassing/. The issue, however, is very subtle as, in order to catch a pokémon, the player does not necessarily need to enter the physical area where the pokemon is placed. A different issue is, however, whether the placing by the game manufacturer of a virtual creature in a private property constitutes an unlawful interference with the right of property.
28. Ryan Calo et al., *Augmented Reality. A Technology and Policy Primer*, Report of the University of Washington Tech Policy Lab (2015): http://techpolicylab.org/wp-content/uploads/2016/02/Augmented_Reality_Primer-TechPolicyLab.pdf
29. Benjamin Freed, "Pokemon Go Locations Are Another Symptom of DC's Demographic Divide," *Washington Post*, August 2, 2016: www.washingtonian.com/2016/08/02/pokemon-go-locations-are-another-symptom-of-dcs-demographic-divide/

30. Alain Labrique, "Pokémon Go! Pandemic or Prescription?" *Global Health Now*, August 2, 2016: www.globalhealthnow.org/2016–08/pokemon-go-pandemic-or-prescription
31. John M. Grohol, "Pokemon Go Reportedly Helping People's Mental Health, Depression," *World of Psychology*, July 11, 2016: http://psychcentral.com/blog/archives/2016/07/11/pokemon-go-reportedly-helping-peoples-mental-health-depression/
32. This would be against the sense of shared community that public common spaces provide as the Pope highlights in LS 151.
33. The Pope borrows this expression from Romano Guardini, a philosopher whose influence in the intellectual vision of the *Laudato Si'* is enormous: Romano Guardini, *Das Ende der Neuzeit: Ein Versuch zur Orientierung* (Würzburg: Im Werkbund-Verlag, 1950), p. 63.
34. A fascinating yet completely unexplored topic is to retrace in AR and similar digital applications elements of the "artificial memory" tradition which saw its golden age in the Renaissance Period and it was systematized in the work of philosopher Giordano Bruno. "Artificial memory," or "the art of memory," described as a set of techniques, habits and mechanical methods comprising images or memory places ("*locii*") as a source of knowledge and a key to understand reality. The memory systems, which are virtual but reflect notable places or physical objects, purport the idea that man can grasp, hold and understand the greater world through visualization and imagination. They aim at providing to the human mind the power to understand the intelligible world beyond worldly appearances. Cf. Frances A. Yates, *The Art of Memory* (London: Routledge and Kegan Paul, 1966), pp. 1–488; Paolo Rossi, *Logic and the Art of Memory: The Quest for a Universal Language* (Chicago, IL: University of Chicago Press, 2000), pp. 1–288.
35. It is fruitful and stimulating at this point to recall that two authors have likened AR to faith that offers to believers a source of an augmented sense of reality: faith which discerns the body of Christ in the host raised by the priest in the holy communion gives an information that the naked eyes do not capture from the reality, causing the perception of a mixed and 'augmented' reality. Cf. Spadaro, *Cybertheology*, p. 110; Peter Phillips and Richard S. Briggs, "The Bible as Augmented Reality: Beginning a Conversation," *Theology and Ministry* 1 (2012): 2.1–10: www.dur.ac.uk/resources/theologyandministry/TheologyandMinistry1_2.pdf